Industrial Application of
Electromagnetic Computer Codes

EURO

COURSES

A series devoted to the publication of courses and educational seminars organized by the Joint Research Centre Ispra, as part of its education and training program.
Published for the Commission of the European Communities, Directorate-General Telecommunications, Information Industries and Innovation, Scientific and Technical Communications Service.

The EUROCOURSES consist of the following subseries:

- Advanced Scientific Techniques
- Chemical and Environmental Science
- Energy Systems and Technology
- Environmental Impact Assessment
- Health Physics and Radiation Protection
- Computer and Information Science
- Mechanical and Materials Science
- Nuclear Science and Technology
- Reliability and Risk Analysis
- Remote Sensing
- Technological Innovation

COMPUTER AND INFORMATION SCIENCE

Volume 1

The publisher will accept continuation orders for this series which may be cancelled at any time and which provide for automatic billing and shipping of each title in the series upon publication. Please write for details.

Industrial Application of Electromagnetic Computer Codes

Edited by

Yves R. Crutzen

Commission of the European Communities,
Joint Research Centre (JRC), Institute for Systems Engineering and Informatics,
Ispra, Italy

Giorgio Molinari

Dipartimento di Ingegneria Elettrica,
Università di Genova, Genova, Italy

and

Guglielmo Rubinacci

Dipartimento di Ingegneria Elettrica,
Università di Napoli "Federico II",
Napoli, Italy

KLUWER ACADEMIC PUBLISHERS
DORDRECHT / BOSTON / LONDON

Based on the lectures given during the Eurocourse on
'Industrial Application of Electromagnetic Computer Codes'
held at the Joint Research Centre Ispra, Italy, November 6–9, 1990

Library of Congress Cataloging-in-Publication Data

```
Industrial application of electromagnetic computer codes / edited by
   Yves R. Crutzen, Giorgio Molinari, Guglielmo Rubinacci.
       p.   cm. -- (Euro courses.  Computer and information science ;
   v. 1)

   1. Electric machinery--Design aand construction--Data processing.
I. Crutzen, Yves R., 1950-   .  II. Molinari, Giorgio, 1941-    .
III. Rubinacci, Guglielmo, 1952-   .  IV. Series.
TK2331.I53   1990
621.3--dc20                                              90-48021
```

ISBN-13: 978-94-010-6799-7   e-ISBN-13: 978-94-009-0697-6
DOI: 10.1007/978-94-009-0697-6

Publication arrangements by
Commission of the European Communities
Directorate-General Telecommunications, Information Industries and Innovation,
Scientific and Technical Communications Service, Luxembourg

EUR 13139
© 1990 ECSC, EEC, EAEC, Brussels and Luxembourg

Softcover reprint of the hardcover 1st edition 1990

LEGAL NOTICE
Neither the Commission of the European Communities nor any person acting on behalf of the
Commission is responsible for the use which might be made of the following information.

Published by Kluwer Academic Publishers,
P.O. Box 17, 3300 AA Dordrecht, The Netherlands.

Kluwer Academic Publishers incorporates the publishing programmes of
D. Reidel, Martinus Nijhoff, Dr W. Junk and MTP Press.

Sold and distributed in the U.S.A. and Canada
by Kluwer Academic Publishers,
101 Philip Drive, Norwell, MA 02061, U.S.A.

In all other countries, sold and distributed
by Kluwer Academic Publishers Group,
P.O. Box 322, 3300 AH Dordrecht, The Netherlands.

Printed on acid-free paper

C O N T E N T S

PREFACE

During the last decade a new generation of software tools has evolved in computational electromagnetics. Both analytical methods and particularly numerical techniques have improved considerably, leading to an extended range of capabilities and an increased applicability of both dedicated and general purpose computer codes.

It is the intention of this volume to review the state of the art in electromagnetic analysis and design, and to describe the fundamentals and the advances in theoretical/numerical approaches coupled with practical solutions for static and time-dependent fields.

In this context, the book illustrates the effectiveness of numerical techniques and associated computer codes in solving real electromagnetic field problems. In addition, it demonstrates the usefulness of modern codes for the analysis of many industrial practical cases. In particular, solutions of magnetostatic and magnetodynamic problems applied to electrical machines, induction heating, non-destructive testing, fusion reactor technology and other industrial are presented and discussed.

The present volume reflects and combines the lectures which are organized in the frame of the Eurocourse programme at JRC Ispra under the sponsorship of the Institute for Systems Engineering and Informatics (ISEI). It is hoped that in this context the Institute and particularly the Systems Engineering & Reliability (SER) Division can play a stimulating role in sponsoring and promoting the diffusion of knowledge in novel areas of computer and information science.

Giuseppe Volta   Robert W. Witty
SER Division Head   ISEI Director

OVERVIEW OF THE 'STATE OF THE ART' IN ELECTROMAGNETIC ANALYSIS AND DESIGN

C.W. TROWBRIDGE
Vector Fields Ltd.
24 Bankside
Kidlington
Oxford OX5 1JE, UK

ABSTRACT. The use of computer software for the design of electromagnetic devices is now almost universal [1,2,3,4,5,6,7]. This lecture presents an overview of the major advances in the subject and attempts to identify the principal limitations. Particular attention is given to the computing environment, 3-D algorithms for electromagnetic field computation and in the use of parallel architectures.

1. Introduction

The widespread use of numerical techniques today demonstrates the great importance of the subject. Efficient solutions can be obtained for a wide range of problems well beyond the scope of analytical methods. The use of numerical methods has overcome the limitations imposed by analytic methods, i.e. their restriction to homogenous, linear and steady-state problems. For example, it is now routine to compute the magnetic forces acting on the components of an electrical machine, taking into account the three dimensional geometry including the slots, conductors, etc. as well as the saturation effects of the material. It is now common practice for engineers to solve many of the complex problems arising during the design phase of sophisticated modern electromagnetic devices by using largely automatic and general purpose procedures; examples include electrical devices of all types as well as the large magnets used in medical diagnostic equipment, particle accelerators, electron beam lenses, fusion magnets and industrial processes—the range of applications is extensive. This success is due largely to the enormous advances that have been made in the power of the digital computer; and although the basis of numerical analysis extends back a long way in history, there is also a steady stream of new developments allowing for more and more difficult problems to be solved. If this activity continues at the present rate it is expected that in the next few years the efficiency of computer methods will reach the stage when genuine computer aided design (CAD) procedures are practical for three dimensional systems, i.e. design *lay-out* software will be integrated with electromagnetic field analysis programs[8].

Numerical analysis is the technique for solving mathematical problems by numerical approximation. By this means solutions are arrived at with the aid of rational numbers and, inevitably, since digital computers are used to perform the computations, only a finite set of

1

Y. R. Crutzen et al. (eds.), Industrial Application of Electromagnetic Computer Codes, 1–27.
© 1990 *ECSC, EEC, EAEC, Brussels and Luxembourg.*

these numbers will be available. In addition to this *rounding error* the size of the computer memory (i.e. the number of *words*) will limit the complexity of the problem that can be solved which in turn means that it will be impossible to achieve a perfect representation of the geometry; thus the actual problem will usually be replaced by a *computer model*, built by the engineer to represent the critical features under investigation. The use of a numerical technique will introduce further *modelling errors* called discretisation or truncation errors arising when the mathematical description, a continuous partial differential equation, is replaced by an approximate numerical description characterised by discrete points.

The discretisation and solution of continuum problems, in this case electrical engineering devices, have been approached differently by mathematicians and engineers, see the text by Zienkiewicz([9] 1977, page 1). The former have concentrated on general techniques, directly applicable to the field equations, such as finite differences approximations and weighted residual procedures, whereas the latter have often used a more intuitive approach exploiting analogues between real discrete elements and finite regions of the continuum. Indeed, electrical engineers have often used circuit analogues to model their problem both experimentally and numerically; and before the widespread use of the digital computer many other analogues were used experimentally such as *resistive paper* and *electrolytic tanks*. The intuitive, direct anology, approach by engineers, particularly in the area of structural mechanics lead to the development of the Finite Element Method [10]; and by 1960 this technique was widely used in other disciplines. It is no accident that these advances came at this time since in the early 1960's occured the rapid development of the digital computer as a universal tool for engineers. In the meantime electrical engineers had, in the main, followed and applied the developments in Finite Differences, now a highly developed discipline of mathematics, and were able to write elegant computer codes particularly for simple static two dimensional configurations with linear media—for example, Hornsby(1967) [11] at CERN developed a successful code used extensively in the design of particle physics magnets.

An important milestone in the solution of electromagnetics field problems came in 1963 with the seminal work of Winslow [12] at the Lawrence Livermore Laboratory California; he developed a discretisation scheme based on an irregular grid of plane triangles, not only by using a generalised finite difference scheme but also by introducing a variational principle which he showed led to the same result. This latter approach can be seen as being equivalent to the Finite Element Method and is accordingly one of the earliest examples of this technique used for electromagnetics. The resulting computer code 'TRIM' [13] and its later descendent 'POISSON' [14] have been used all over the world. Finite difference techniques continued to be applied by electrical engineers throughout the 1960's and early 1970's, notably the work of Trutt at Delaware [15], Erdelyi *et al* at Colorado [16], and Viviani [17] *et al* in Genoa, and in three dimensions by Muller and Wolf at AEG Telefunken, Germany [18].

However by the early 1970's the Finite Element Method was under scrutiny by the mathematicians and substantial generalisations were made [19] and many cross links were established with earlier work on variational calculus and generalised weighted residuals [20]. The important advantages of finite elements over finite differences were being exploited, i.e. the ease of modelling complicated boundaries and the extendibility to higher order approximations and in 1970 came the first application of the method to rotational electrical machines by Chari and Silvester [21]. From this time on the use of the method became widespread leading to generalised applications for time dependent and three dimensional problems by the group at Rutherford Laboratory with the production of the codes PE2D and TOSCA [22,23].

A parallel development to the above has been with integral methods; these integral formulations, unlike differential formulations which solve the defining partial differential equations (e.g. Poisson's Equation), use the corresponding integral equation forms, e.g. equations based on Gauss' theorem. The moment method is an example of an integral formulation, see the text by Harrington (1968) [24] for a basic treatment; yet another class of an integral procedures are the so called boundary element methods [25,26] based on applications of Greens integral theorems. Whilst these methods are often difficult to apply they can produce accurate economic solutions and have been used extensively in certain static and high frequency problems, an example of a general purpose program, first developed in 1971, is the magnetisation integral equation code GFUN which was specifically designed for three dimensional static problems [27].

In this lecture the author gives a personal view on the present status of electromagnetic computing. The following topics will be considered:

- Evolution of Code Development

- Limitations in Contemporary Codes Unsolved Problems

- A Modern Computing Environment CAD system for the designer

- Economics of 3-D Computation

- 3-D Field Formulations

- Impact of Parallel Processing

- The Way Ahead

2. Evolution of Code Development

It is instructive to see over the years how the frontiers of tractable problems have been extended and to observe what ground still needs to be covered. In Table 1 are listed the dates when codes with the indicated functionality became generally available. It will be noticed that there was a significant use of integral equation techniques in the early days leading to both two and three dimensional codes. This development was curtailed mainly due to the excessive computing power needed to solve systems of equations whose matrix of coefficients is fully populated. This subject will be returned to later when parallel computers are discussed. The finite element method used to solve the differential forms of the field equations has been strikingly sucessful leading to general purpose software. Thus by the year 1990 3-D solutions can be achieved by electromagnetic device designers in industry for a fairly large range of problems.

Time Evolution of Code Development							
Function	1960	1965	1970	1975	1980	1985	1990
Statics	D2Df I2D	D2D I3D	D2Dnp I3Dn	D2Da	D3Dna	D2Dse	D2Dv
Steady State A.C.			D2D I2D	D2Dn*	D3D†	D2De	D3D
Transient			I2D	D2D	D2Dn	D3D†	D3D
Full Maxwell						D2D	D3D
Motion					D2D‡	D3D‡	D3D
Coupled Problems				D2Df	D2D		D3D

Key

* Approximate model;   ‡Uni-directional velocity;   †Restricted formulation

D2D Differential 2-D	n...non-linear	a...anisotropy
D3D Differential 3-D	p...permanent magnet	f...finite difference
I2D Integral 2-D	s...sacalar hysteresis	v...vector hysteresis
I3D Integral 3-D	e...error analysis	

Table 1: *The Evolution of Fuctionality in EM Codes*

3. Limitations in Contemporary Codes

Although the functionality of contemporary codes has grown over the years, nevertheless as the functionality increases so do the requirements of users. The following list outlines areas where considerable work is still required:–

- Vector Hysteresis

  Major criticisms of numerical methods have been made because, far too often, inadequate material models are used. Considerable research is currently underway to develop better material models, including hysteresis effects [28,29,30]. Nevertheless the sheer complexity involved in keeping track of minor loops remains a daunting task.

- Far Field

  A major consideration in any electromagnetic field analysis is the placement of the far-field boundary. In many cases the natural boundary of a magnet is essentially at infinity although in practice the presence of remote objects and their possible affect on the field shape will have to be taken into account. Special care will be needed with all methods based on solving the differential form of the defining equations. A number of approaches to this problem have been described in the literature [31]:

  1. Terminating the field at a sufficiently large distance.

     This of course begs the question and at the very least will require several solutions in order to achieve confidence. A useful technique here is to solve two problems at each trial with Dirichlet and Neumann boundary conditions respectively in order to bound the solution.

  2. To use special finite element basis functions which have the correct assymptotic behaviour for large distances[32]. The major advantage of using special finite elements is that the matrix size and bandwidth are not significantly increased, but on the other hand some knowlege of how the field decays is needed in order to fix scaling parameters.This approach has been sucessfully implemented in a number of cases [31].

  3. A number of methods based on the idea of a super global element to model the exterior region. The method of recursive ballooning [33] in which the global element is generated iteratively by succesively adding concentric rings of *scaled elements* to embrace the finite element model. At each step the adjacent nodes between rings are removed so there are only nodes on the original and far boundaries. Boundary conditions at the far boundary can now be applied. The method is not so effective in 3-D.

     A major difficulty in these global element methods is that the matrix structure is strongly affected and that the band-width will increase to a point where the matrix becomes essentially full and an advantage of the differential approach disappears.

  4. Mapping techniques have also been extensively used to transform the exterior infinite space to a finite space. The classical Kelvin transformation (see Kellogg [34], page 232) is an example of this in which the transformation $rr' = a^2$

where r and r' are the inverse points with respect to a sphere of radius a. An inversion in a sphere is one-to-one except that the centre of the sphere of inversion has no corresponding point. The neighbourhood of the origin maps into a set of points at a large distance—into an infinite domain. This transformation has been used in a number of finite element systems to model the infinite domain in which the exterior space to a sphere surrounding the actual model is solved as an interior problem by means of the Kelvin transformation. The nodes of the two spaces, now bounded by spheres, are connected by forcing their solution values to be identical [35], [36], [37].

5. The need for special methods is obviated totally if integral methods are used which is one of the major advantages of this approach and can be recommended for small to medium sized problem or larger if computers with parallel architectures are available.

- 3D Field Formulations
  This fundamental topic was discussed in the previous section and remains the most important single issue in computational electromagnetics.

- Moving Boundaries
  The analysis of moving sytems with a high magnetic *Reynold's Number* can lead to singularly perturbed problems. Thus the standard Poisson type equations are modified to the form, $\epsilon\nabla^2\phi + v\mu\sigma\frac{\partial\phi}{\partial x} = f$, where $0 < \epsilon \ll 1$. This can cause instability in the solution if standard finite element methods are used (see reference [38], page 40). Considerable success with problems of this type has been achieved in the area of computational fluid mechanics which has carried over to electromagnetics [39,40]. At this point it may be pertinent to ask whether integral methods may offer significant advantages here since free-space between moving conducors would not require meshing!

- Coupled Problems
  Real industrial problems involve other technologies in addition to electromagnetics—e.g. thermal, structural and fluid. There are many inportant questions to be discussed in this context. One such is to what extent can solving the separate systems in series be used, i.e. using sequential algorithms which minimise the size of the solution space but may, in practice, fail to converge. Or, alternatively accept the enormous costs of parallel, fully coupled analysis. There is a growing body of significant work on coupled phenomena reported in the literature, see reference [7], pages 536-582.

- Optimisation
  It is, of course, the engineer's proper role to provide the creativity and not to waste too much time on *what if?* experiments or, at least, he should consider if an algorithm can be devised in order to provide answers rather than *knob turning* which misses or could never reach a solution in time—the number of states for a binary choice escalates as 2^n where n is a number of variables.

Put another way, the process of design often requires the determination of boundary values that produce a desired field, i.e. it is the solution to the inverse problem that is needed. As an example consider the problem of determining a boundary shape to produce a uniform field in the gap of an electromagnet. In the region under the pole, the magnetic field will, in particular, depend upon the geometry, material and current sources. In particular, if shims are to be used, the optimisation problem is to determine the values of their width and depth that produce a constant value of B_p inside a specified domain. One way of doing this is to solve the field equations by finite elements or boundary elements. To do this suppose only the two limiting values of each shim parameter are explored, i.e. for two shims there will be 8 states, thus there will be $2^8 = 256$ cases to run! This clearly is unfair because trends will be observed on the way and experience will eliminate many of the trials, never the less a lot of computation and time is to be expected. In cases like this automatic optimisation should be considered; there are several methods reported in the literature including least square techiques [27,41], evolution strategies [42,43] and simulated annealing [44]. The near hysterical *hype* of artificial intelligence is also having an impact on electromagnetics design but at present this remains shallow and suspect. Nevertheless, the related and practical procedures using knowledge base methods and *expert systems* has already influenced new research projects in integrated design analysis [8].

4. A Modern Computing Environment

4.1. THE DESIGN PROCESS

Not only have there been advances in the functionality of numerical codes but also the computing environment that these codes work in has changed beyond recognition from the days of the mainframe and batch processing. In Figure 1 is shown the normal procedure of using a field code for design. The designer iterates toward his design solution by a more or less heuristic method(*cut and try?*) supplemented by the engineers creativity and experience. In order to carry out this process efficiently the software system has three usual stages of pre-processing, solution processing and post-processing.

4.2. A CAD SYSTEM FOR DESIGN

In a modern environment the user should expect the following facilities:

- Fast Solutions

- Error Estimation

- Practical Ergonomics for Data Input

- Meaningful Post-Processing

- Automatic Discretisation

8

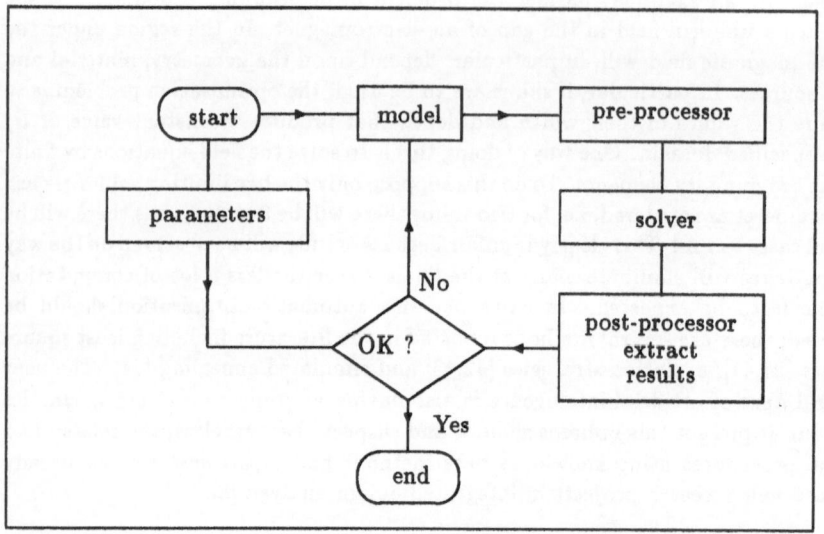

Figure 1: *Flow Diagram for Heuristic Design*

Thus the following components are needed:

- Data Base

- Pre-Processor

- Adaptive Solver

- Post-Processor

- Knowledge Base

In Figure 2 is shown two versions of a design environment. Figure 2(a) shows the normal system in which the pre-processor includes data input, model building and mesh generation. User controlled meshing is extremely tedious. Although fully automated meshing is now a practical possibility it needs to be combined with error estimation in order to allow the generation of optimal meshes. This approach is now becoming quite common for 2-D systems [45,46] and can be expected in 3-D systems before long. Figure 2(b) shows a more ideal system in which the solution processor includes an adaptive mesh generator controlled by *a posteriori* error estimation [45].

5. Economics of 3D Field Computation

Despite advances in solution techniques and computer hardware 3D computation is prohibitive for many applications.This is mainly due to additional complexity involved in moving from 2-D to 3-D. There are two main aspects:

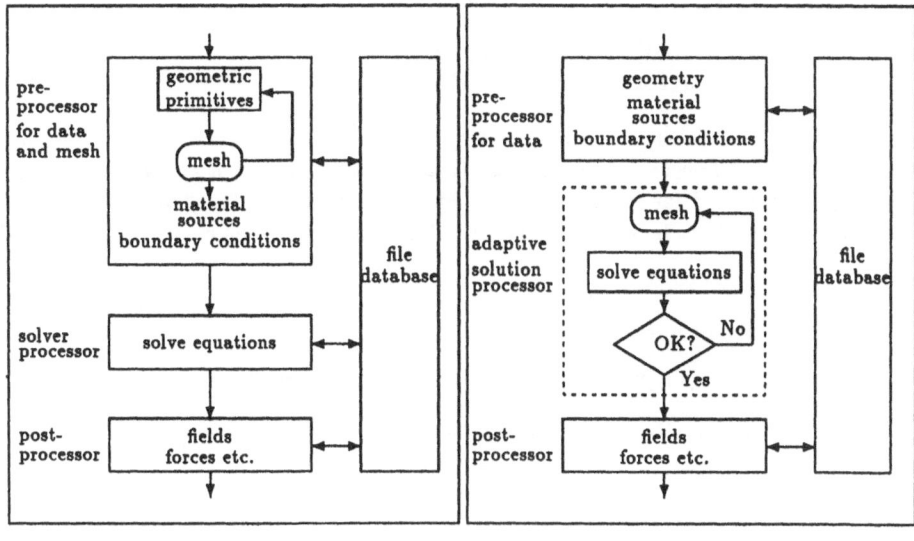

(a) Normal System   (b) Ideal System

Figure 2: *Components of CAD System*

1. The 3D geometry itself leads to a larger sytem matrix and and to a far more complex mesh generation task.

2. Vector nature of the field for time dependent problems causes a ≥ 3 fold increase in the system matrix size.

This has a dramatic effect on the solver compute time T, i.e.

$$T = \alpha n^a + \beta n^b + \gamma n^c f(n)$$

where the three terms refer to the source field, matrix set-up, and matrix solution times respectively. The machine dependent constants usually satisfy:

$$\alpha \gg \beta \gg \gamma$$

Consider two examples:

1. Differential Operator
   using a preconditioned conjugate gradient method,e.g. ICCG
   $a = 1, b = 1, c = 0, f \equiv nlog(n)$
   i.e. for a problem with 10×10 nodes in xy plane:

2D Scalar	3D Vector	nlog(n) ratio
10x10	3x10x10x10	~ 50

2. Integral Operator using Gaussian Ellimination, then $a = 1, b = 2, c = 3, f \equiv 1$

2D Scalar	3D Vector	n^3 ratio
10x10	3x10x10x10	~ 27000

The above examples clearly demomstrate why the differential method has been the most sucessful to date!. However to be fair *like is not being compared with like* since integral equation solutions need far fewer unknowns. Furthermore if parallel computers are used a significant speed-up is to be expected since the integral formulation is intrinsically parallel (see section 8.).

6.   3-D Field Formulations

6.1.   QUASI-STATIC ELECTROMAGNETIC FIELD EQUATIONS

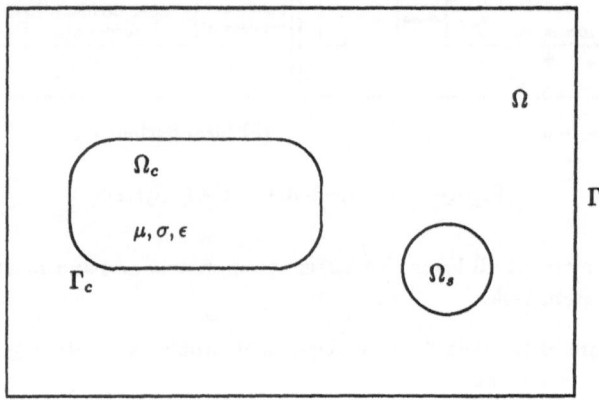

Figure 3: *Simple Model Configuration for Eddy Currents*

The basic equations describing electromagnetic fields are listed here, without detailed explanations, to remind the reader of the concepts and nomenclature. For the purpose of introducing the field equations it is convenient to consider the elementary model problem shown in Figure (3) in which a volume of conducting material Ω_c, with magnetic permeability μ and electrical conductivity σ, bounded by a surface Γ_c, is contained within a global volume of free space Ω bounded by a surface Γ which, furthermore, may be extended to infinity if required. The global region may also contain a number of prescribed conductor sources Ω_s which do not intersect Ω_c. This configuration arises in a very large number of applications of practical importance in industry and problems of this type have been the starting point for many developers of computer algorithms. However care is needed if Ω contains multiply connected regions.

It is known that if the dimensions of the regions Ω_c and Ω_s are small compared with the wavelength of the prescribed fields then the displacement current term in Maxwell's equations will be small compared to the free current density $\mathbf{J}$ and there will be, essentially, no radiation (see Stratton, page 277) [47]. This regime means a return to the pre-Maxwell field equations, the so called quasi-static case, where Ampere's law is a good approximation. In this situation the field equations can be approximated by:

$$\nabla \cdot \mathbf{D} = \rho \quad (Gauss's\ Law) \tag{1}$$

$$\nabla \cdot \mathbf{B} = 0 \tag{2}$$

$$\nabla \times \mathbf{E} = -\frac{\partial \mathbf{B}}{\partial t} \quad (Faraday's\ Law) \tag{3}$$

$$\nabla \times \mathbf{H} = \mathbf{J} \quad (Ampere's\ Law) \tag{4}$$

where $\mathbf{D,B,E,H}$ are the usual field vectors, ρ and $\mathbf{J}$ the free charge and current densities respectively [47]. The field vectors are not independent since they are further related by the material constitutive properties;

$$\mathbf{D} = \epsilon \mathbf{E} \tag{5}$$

$$\mathbf{B} = \mu \mathbf{H} \tag{6}$$

where ϵ and μ are the material permittivity and permeability respectively. The current density in a conductor moving with relative velocity $\mathbf{v}$ is generated by the Lorentz force and is given by:

$$\mathbf{J} = \sigma(\mathbf{E} + \mathbf{v} \times \mathbf{B}) \quad (Ohm's\ Law) \tag{7}$$

where σ is the material conductivity. In practice μ and σ may often be field dependent quantities, and furthermore, some materials will exhibit both anisotropic and hysteretic effects. The current continuity condition follows from Eq. (4), and is:

$$\nabla \cdot \mathbf{J} = 0. \tag{8}$$

The four field vectors must satisfy the following conditions at the interfaces between regions of different material properties;

$$(\mathbf{B_2} - \mathbf{B_1}) \cdot \mathbf{n} = 0 \tag{9}$$

$$(\mathbf{D_2} - \mathbf{D_1}) \cdot \mathbf{n} = \omega \tag{10}$$

$$(\mathbf{H_2} - \mathbf{H_1}) \times \mathbf{n} = \mathbf{K} \tag{11}$$

$$(\mathbf{E_2} - \mathbf{E_1}) \times \mathbf{n} = 0 \tag{12}$$

where $\mathbf{K}$ and ω are the surface current and charge densities respectively. These relations follow directly from the limiting forms of field equations, Equations (1) to (4), applied at the interfaces.Furthermore, for the quasi-static case, Eq. (8) implies continuity of the normal component of current density at material interfaces.

6.2. MAGNETIC VECTOR POTENTIAL A

Since the field vector $\mathbf{B}$ satisfies a zero divergence condition, it can be expressed in terms of a vector potential $\mathbf{A}$ as follows:

$$\mathbf{B} = \nabla \times \mathbf{A}, \tag{13}$$

and then, from the field equations, Eq. (3), it follows that,

$$\nabla \times (\mathbf{E} - \frac{\partial}{\partial t}\mathbf{A}) = 0, \tag{14}$$

and hence by integrating to give,

$$\mathbf{E} = -(\frac{\partial}{\partial t}\mathbf{A} + \nabla V), \tag{15}$$

where V is a scalar potential. Neither $\mathbf{A}$ nor V are completely defined since the gradient of an arbitrary scalar function can be added to $\mathbf{A}$ and the time derivative of the same function can be subtracted from V without affecting the physical quantities $\mathbf{E}$ and $\mathbf{B}$. These changes to $\mathbf{A}$ and V are the so called gauge transformations, and uniqueness is usually ensured by specifying the divergence (gauge) of $\mathbf{A}$ and sufficient boundary conditions. Thus in region Ω_c the field equations in terms of $\mathbf{A}$ and V are as follows:

$$\nabla \times \frac{1}{\mu}\nabla \times \mathbf{A} + \sigma(\frac{\partial \mathbf{A}}{\partial t} + \nabla V) = \mathbf{J}, \tag{16}$$

$$\nabla \cdot \sigma(\frac{\partial \mathbf{A}}{\partial t} + \nabla V) = 0 \tag{17}$$

and, in the global region where $\sigma = 0$ and $\nabla \times \mathbf{H} = \mathbf{J_s}$, reduces to

$$\nabla \cdot \mu\nabla\phi = 0, \tag{18}$$

where ϕ is the reduced magnetic scalar potential with $\mathbf{H} = \mathbf{H_s} - \nabla\phi$ for a source field $\mathbf{H_s}$.

At points just inside a conductor the continuity conditions imply

$$\mathbf{J_n} = -\sigma(\frac{\partial \mathbf{A}}{\partial t} + \nabla \mathbf{V}) \cdot \mathbf{n} = 0 \tag{19}$$

$$\frac{\partial \mathbf{A_n}}{\partial t} + \frac{\partial V}{\partial n} = 0 \tag{20}$$

at conductor surfaces, and at interfaces, across which the conductivity changes from σ_1 to σ_2 implies that,

$$\sigma_1(\frac{\partial}{\partial t}\mathbf{A_1} + \nabla V_1) \cdot \mathbf{n} = \sigma_2(\frac{\partial}{\partial t}\mathbf{A_2} + \nabla V_2) \cdot \mathbf{n}. \tag{21}$$

Are Equations (16) and (17) sufficient? It is clear that Eq. (17) is a consequence of taking the divergence of Eq. (16) and is not, therefore, independent. Some investigators [48,49] have obtained unique solutions to Equations (16) and (17), as they stand, but they show that the uniqueness depends upon the particular numerical procedures used. This leaves a flexibility of the system unused thus is it possible to use this flexibility to advantage? In any case it is necessary to specify the divergence (gauge) of $\mathbf{A}$ and appropriate boundary conditions to ensure uniqueness. The two commonest conditions are the Coulomb ($\nabla \cdot \mathbf{A} = 0$), and the Lorentz ($\nabla \cdot \mathbf{A} = -\mu\sigma V$) gauges.

(a) Coulomb Gauge [50,51]

For the Coulomb gauge, since $\nabla \cdot \mathbf{A} = 0$

$$\nabla \times \frac{1}{\mu}\nabla \times \mathbf{A} - \nabla\frac{1}{\mu}\nabla \cdot \mathbf{A} \;=\; -\sigma(\frac{\partial}{\partial t}\mathbf{A} + \nabla V) \tag{22}$$

$$\nabla \cdot \sigma\frac{\partial \mathbf{A}}{\partial t} + \nabla \cdot \sigma\nabla V \;=\; 0. \tag{23}$$

Now, unlike before, Eq. (23) will not be obtained by taking the divergence of Eq. (22). In this case,

$$\nabla^2(\frac{1}{\mu}\nabla \cdot \mathbf{A}) = 0 \tag{24}$$

and Eq. (23) is independent. Furthermore, from Eq. (24) it can be seen that if

$$\mathbf{A} \cdot \mathbf{n} = 0 \tag{25}$$

is prescibed on the conductor boundary, then the Coulomb gauge is enforced. However there will be practical difficulties in imposing Eq. (25) in general. For the case where σ is constant, Eq. (23) reduces to $\nabla^2 V = 0$, and since Eq. (25) is true on the conductor boundary then, from Eq. (20), V at best must be a constant.

(b) Lorentz Gauge [52,53]

The governing equations for $\mathbf{A}$ with no gauge applied are given by Equations (16) and (17) which can be re-written as

$$\nabla \times \frac{1}{\mu}\nabla \times \mathbf{A} \;=\; -\sigma(\frac{\partial}{\partial t}\mathbf{A} + \nabla V) \tag{26}$$

$$\nabla \cdot \sigma\frac{\partial \mathbf{A}}{\partial t} + \nabla \cdot \sigma\nabla V \;=\; 0. \tag{27}$$

As already stated Eq. (27) follows from Eq. (26) by taking the divergence and it does, in fact, confirm that $\nabla \cdot \mathbf{J} = 0$, and thus this equation is not independent. The Lorentz Gauge condition can now be used to obtain a sufficient set of defining equations with $\mathbf{A}$ and V separated. The low frequency form of the Lorentz gauge is,

$$\nabla \cdot \mathbf{A} = -\mu\sigma V. \tag{28}$$

The defining Equations (26) and (27) permit a number of ways of implementing the Lorentz gauge [52]. One such is to to simply replace ∇V in Eq. (26) from Eq. (28). Another is to to replace ∇V in Eq. (26) and $\nabla \cdot \frac{\partial \mathbf{A}}{\partial t}$ in Eq. (27) from Eq. (28). Variant one has the disadvantage of producing an asymmetric algebraic system of equations when the equations are discretised by the finite element method, which, at the very least, will require more computer memory and other side effects. The second approach permits a symmetric (minimal memory) system to be constructed by substituting

$$v = \int_0^t V(s)ds. \tag{29}$$

The second variant produces the following two equations:

$$\nabla \times \frac{1}{\mu}\nabla \times \mathbf{A} = -\sigma\frac{\partial \mathbf{A}}{\partial t} + \sigma\nabla(\frac{1}{\mu\sigma}\nabla \cdot \mathbf{A}) \tag{30}$$

$$-\mu\sigma^2\frac{\partial V}{\partial t} + \nabla \cdot \sigma\nabla V = 0. \tag{31}$$

where σ is assumed piecewise constant. Thus this approach reflects the classical motivation for the Lorentz Gauge namely, to decouple the vector potential from the scalar potential. These equations produce a unique solution if V or $\mathbf{A} \cdot \mathbf{n}$ is given on conductor surfaces for non-zero frequency and $\mathbf{A} \cdot \mathbf{n}$ is given on conductor surfaces for zero frequency [52,53]. As with the Coulomb gauge the imposition of $\mathbf{A} \cdot \mathbf{n}$ is not generally straightforward. Recent work has shown that

$$\mathbf{A} \cdot \mathbf{n} = \beta\sigma\mathbf{V}$$

where β is a suitably chosen constant is sufficient to guarantee uniqueness from DC to Daylight [54].

(c) Modified Vector Potential A or E

Some investigators [55,56,57,58,59] have eliminated the scalar potential V completely by specifying

$$\frac{\partial \mathbf{A}^*}{\partial t} = \frac{\partial \mathbf{A}}{\partial t} + \nabla V \tag{32}$$

where $\mathbf{A}^*$ is termed the modified vector potential, however this can now be seen as a consequence of the Coulomb gauge above. It should also be pointed out that this formulation is essentially equivalent using the electric field vector $\mathbf{E}$ [56]. Therefore if the modified Vector potential is used the field equations reduce to:

$$\nabla \times \frac{1}{\mu}\nabla \times \mathbf{A}^* + \sigma\frac{\partial \mathbf{A}^*}{\partial t} = \mathbf{J}, \tag{33}$$

The modified vector potential (Eq. (32)) is equivalent to the Gauge condition $\nabla \cdot \sigma\mathbf{A}^* = 0$ which for a single conductor region with constant σ reduces to the Coulomb gauge $\nabla \cdot \mathbf{A}^* = 0$. The difference between these two choices seems insignificant but in fact is profound. The first equation produces a continuity condition

$$\sigma_1\mathbf{A}_1^* \cdot \mathbf{n} = \sigma_2\mathbf{A}_2^* \cdot \mathbf{n} \tag{34}$$

whereas the second equation gives,

$$\mathbf{A}_1^* \cdot \mathbf{n} = \mathbf{A}_2^* \cdot \mathbf{n} \tag{35}$$

From these two relationships it becomes clear immediately that condition (34) is going to introduce difficulties when two conductor regions of differing conductivities abutt. Note that formally the modified vector potential (Eq. (32)) approach does not fail to handle this situation but gives rise to a level of complexity judged by many implementers to be unacceptable. The other major weakness of the modified vector potential approach is that the formulation is based on the **curl curl** operator which can produce a singular matrix in its numerical form when σ tends to zero. The quality of the solution therefore resides in some sense with the right hand side term which in turn depends on the frequency and electrical conductivity. Practically the implications are, that for low frequency problems (e.g. 1Hz) the system matrix becomes ill conditioned. So far as more general problems are concerned, i.e. those involving a variation in conductivity or when solutions are required over frequencies down to zero, the gauged systems (a) and (b) above are to be preferred. Never-the-less for a substantial range of problems the $\mathbf{A}^*$ method has produced good results [56,59].

6.3. MAGNETIC FIELD VECTOR H

The corresponding field equation in terms of the field intensity **H** is obtained directly from the field equations (see lecture on differential and integral formulations) and is given by:

$$\nabla \times \nabla \times \mathbf{H} + \sigma \frac{\partial}{\partial t}(\mu \mathbf{H}) = 0 \tag{36}$$

however in the global region where $\sigma = 0$ and $\nabla \times \mathbf{H} = 0$, equation (36) limits to a Laplacian field in ϕ. A direct application of clasical finite elements, using nodal basis functions, in the conducting regions will lead to difficuties. This is because the physical interface conditions do not emerge naturally when a Galerkin weighted residual method is applied to equation (36). The solution to this problem is to use the edge variable elements which enforce the correct interface conditions [60,61]. Edge variables for tetrahedral elements are defined by

$$\mathbf{h(r)} = \mathbf{a} + \mathbf{b} \times \mathbf{r} \tag{37}$$

where **r** is the position vector and **a** and **b**, respectively, are vectors dependent on the geometry of the element. The basis function expansion is the given by

$$\mathbf{H} = \sum \mathbf{h_e(r)H_e} \tag{38}$$

where $\mathbf{h_e}$ is the vector basis function for edge e, and $\mathbf{H_e}$ is the value of the field along an element edge, see Figure 6.3.. The functions, equation (37) have the following useful properties, $\nabla \cdot \mathbf{h} = 0$ and $\nabla \times \mathbf{h} = 2\mathbf{b}$.

Most importantly, the edge variable ensures that the tangential component of **H** is continuous whilst allowing for the possibility of a discontinuity in the normal component. In non-conducting regions where the field is modelled by a scalar potential, standard nodal elements can be used, see Figure 6.3.. At the interface between conducting and non-conducting regions the nodal basis functions couple elegantly to the edge variables [60], i.e.

$$h_1 = \frac{(\phi_2 - \phi_1)}{l}$$

In an important implementation of this method [61] the scalar region has been solved using a boundary integral equation allowing the infinite space to be modelled.

16

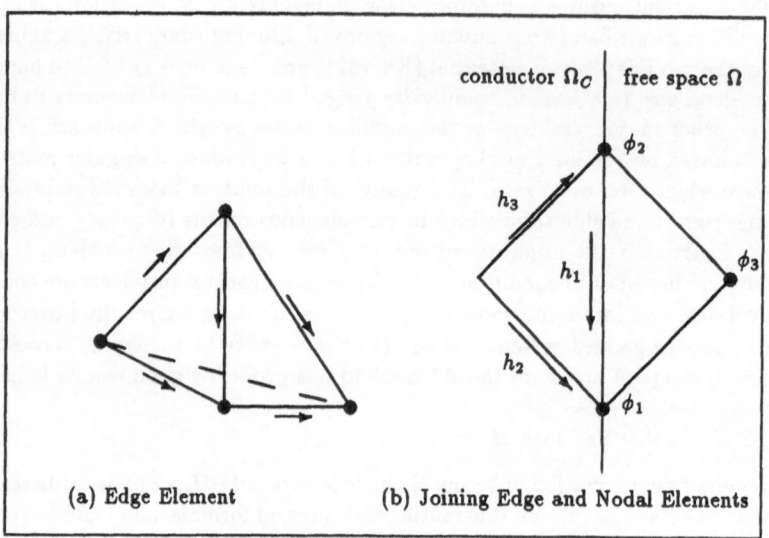

(a) Edge Element   (b) Joining Edge and Nodal Elements

Figure 4: *Edge and Nodal elements*

6.4. ELECTRIC VECTOR POTENTIAL T

Since the current density **J** satisfies the divergence condition, equation, it can be expressed in terms of a vector potential **T** as follows:

$$\mathbf{J} = \nabla \times \mathbf{T}, \tag{39}$$

with

$$\mathbf{H} = (\mathbf{T} - \nabla\phi), \tag{40}$$

where ϕ is a magnetic scalar potential. The corresponding field equations in terms of the electric vector potential **T** are obtained from the field equations (1) to (4) and are given by

$$\nabla \times \frac{1}{\sigma} \nabla \times \mathbf{T} + \mu \frac{\partial}{\partial t}(\mathbf{T} - \nabla\phi) = 0 \tag{41}$$

and,

$$\nabla \cdot \mu(\mathbf{T} - \nabla\phi) = 0 \tag{42}$$

for the case where σ is piecewise constant, and in the global region where $\sigma = 0$ and $\nabla \times \mathbf{H} = 0$, equation (42) limits to scalar potential region as before. There have been a number of successful implementations of this method [62,63,64] using conventional finite elements, these include also treatment of ferro-magnetic regions [62,65] and in this context a recent paper introduces an associated **T** vector to represent magnetisation currents [64]. Some investigations have not dealt fully with the problem of gauge or continuity. Now equation (40) is not unique with respect to **T** since the addition of a gradient of a scalar potential does not affect **J**, cf. a similar situation with the magnetic vector potential. A suitable gauge is the following

$$\mathbf{T} \cdot \mathbf{u} = 0 \tag{43}$$

where u is an arbitrary vector field which does not posssess closed field lines [66,63]. The enforcement of equation (43) not only allows **T** to be constrained but enables the number of degrees of freedom to be reduced from three to two. This method, with the gauge equation (43), has been implemented successfully in both differential and integral form using the edge variable element approach to ensure natural continuity [63,67]. However, the formulation needs extending if ferro-magnetic regions are to be included.

7. Summary of Differential Formulations

7.1. WHICH VECTOR?

Unlike the 2-D case there is no predominant single formulation for 3-D Eddy current computations. The principal formulations are summarized in table 2 and as the literature makes clear each has advantages and weaknesses depending upon context. There are many ommisions. The criterion for inclusion was to include techniques capable of some degree of generality. For example, in 2-D robust procedures are available in computer codes. Similarly 3-D non-linear magnetostatics problems can be handled by robust codes with efficient pre and post-processing [68].

The situation with 3-D eddy currents is not as stable with only rather restricted functionality. Using special methods for specific problems, (e.g. integral methods based on circuit analogues [69,70,71]) is often helpful, and it is interesting that a sharper understanding of the connection between these approaches and methods described here is emerging [61,67].

7.2. BENCHMARKING

A significant event in recent has been the establishment of the international workshops on eddy current code validation (TEAM). Six bench-mark problems have been defined and solved by a mix of special and general methods. The results of the first set have now been published [72], a further group of more advanced problems are underway.

Researchers are using these problems to help in developing new methods as well as to validate existing computer codes.
Which method is best?

   a Good results obtained for a wide range of general methods and special purpose methods with limited functionality.

   b Special methods often give good results more economically, but general methods that allow a reduction in the number of components are sometimes more efficient.

   c The development of special methods needs expensive resources, and they often have no life beyond their original purpose. To minimise this investment special methods should be integrated into general purpose pre and post-processing environments.

   d General methods are needed by designers who have not the resources to develop special techniques.

Eddy Current Field Formulations		
Method	Governing Equation	Gauge Condition
$\mathbf{H} - \phi$	$\nabla \times \nabla \times \mathbf{H} = -\mu\sigma\frac{\partial \mathbf{H}}{\partial t}$	$\nabla \cdot \mathbf{H} = 0$
$\mathbf{A} - V - \phi$	$\nabla \times \frac{1}{\mu}\nabla \times \mathbf{A} = -\sigma(\frac{\partial}{\partial t}\mathbf{A} + \nabla V)$ $\nabla \cdot \sigma(\frac{\partial}{\partial t}\mathbf{A} + \nabla V) = 0$	$\nabla \cdot \mathbf{A} = 0$ $\nabla \cdot \mathbf{A} = -\mu\sigma V$
$\mathbf{A}^* - \phi$	$\nabla \times \frac{1}{\mu}\nabla \times \mathbf{A}^* = -\sigma\frac{\partial \mathbf{A}^*}{\partial t}$	$\nabla \cdot \sigma\mathbf{A} = 0$
$\mathbf{T} - \phi$	$\nabla \times \frac{1}{\sigma}\nabla \times \mathbf{T} = -\mu\frac{\partial}{\partial t}(\mathbf{T} - \nabla\phi)$ $\nabla \cdot \mu(\mathbf{T} - \nabla\phi) = 0$	$\mathbf{T} \cdot \mathbf{u} = 0$

NOTES

Free Space: In all cases ϕ satisfies $\nabla^2\phi = 0$

Conductors:
H
      Gauge implied by formulation
      Edge Elements appropriate in conductor regions
      Nodal elements in free-space regions

A – V   Gauge needs to be applied, Lorentz or Coulomb?
A*   Gauge implied by the formulation

T – ϕ Gauge applied to reduce numer of components

Table 2: *Field Vectors for Eddy Currents*

Equation Solution Time(Seconds) for One Iteration							
Nodes (No Eqns)	μVax II	SUN 3/140	FPS M64/60	STELLAR GS100	ARDENT TITAN	CYBER 990	CRAY XMP-48
849		77		11			
11305				141	189	19	13
18700	8826		224				
34000	13320						72

Table 3: *Performance figures for the TOSCA Code*

8. The Impact of Parallel Processing

Parallel computing heralds a new vista for computational mechanics, or so it is said, but do the results bear any resemblance to the propaganda? The architecture of computers has played a decisive role in defining their performance—there has been a speed-up of 10^6 in floating point computations over the last three decades but only a factor of ~ 150 can be attributed to improvements in technology [73]. From the user viewpoint however, the number of architecture types is bewildering and the classification schemes used are often confusing. The present generation of parallel computers appear to be based on three main architectural characteristics: memory structure, *grain size* and control. Thus memory can either be distributed, with each processor controlling its own memory, or it can be shared among several (all) processors. Grain size refers to the computational power of each node, i.e. each processor and its associated memory. If each node is capable of holding an entire program and has the order of a megabyte of store, then the grain size is said to be large, and conversely if nodes have a single bit processor and no more than a few kilobytes of store then the grain size is said to be small. Control can either be SIMD (Single Instruction, Multiple Data) in which the processors execute identical instructions in *lock-step*, or MIMD (Multiple...) in which the processors run different codes asynchronoulsly. Examples are: (a) distributed memory, small grain size, SIMD machines like the Digital Array Processors(DAP's), (b) shared memory, small grain size SIMD machines like the family of vector procssors, (c) shared memory, large grained, MIMD machines like the CRAY-2, and (d) distributed memory, large grained MIMD machines like the INMOS transputer.

Table 3 shows a somewhat limited set of performance results for the electromagntic code TOSCA [74] run on several machines [75,76]. These results were obtained without the necessary re-coding to fully exploit the machines, except in the case of the CRAY implementations in which the system matrix was scanned along the diagonals to minimise the number of short vectors. It can be seen that immediate gains in performance are possible with the existing software, however further improvements should be possible with code optimization and by the selection of algorithms which match the architecture [77]. The TOSCA code is based on a differential method and uses linear algebra which is optimal for serial computers; consequently, codes of this type will not perform optimally on type (d) multiproceessing MIMD machines like the transputer because their algorithms are not intrinsically parallel. With integral equations the situation is very different.

8.1. INTEGRAL METHODS AND TRANSPUTERS

The use of Integral methods for solving electromagnetic field problems has always had many advocates. Indeed, one of the first computer codes for 3D non-linear magnetostatics to have an existence beyond its original purpose was based on an integral equation solved by applying the method of moments [27]. The advantages of integral formulations compared to the standard differential approach using finite elements are: (a) only active regions need to be discretised, (b) the far-field boundary condition is automatically taken into account, and (c) the fields recovered from the solution are usually very smooth. Unfortunately, the computational costs are high and rapidly escalate as the problem sizes are increased. For example, the time T to compute a complete solution can be written, $T = \alpha n + \beta n^2 + \gamma n^3$, where α, β, and γ are context dependent constants governing the source computation, matrix set-up and solution process respectively, and n is the number of unknowns. It usually happens that long before the n^3 term dominates the other two terms are limiting the size of problem that can be handled for a given computer. However, these terms are essentially parallel operations, and furthermore, a significant degree of parallelism can be exploited in the solution process itself. A typical example of an integral equation procedure arises when boundary elements are used to solve electromagnetic problems. This will be illustrated by some results from a study ing using a transputer enhanced workstation [78]. The transputer, developed and manufactured by the INMOS Ltd. [79], is a complete computer on a single chip. This boundary element method is at the heart of BIM2D [74], which solves problems of the type illustrated in Figure 5(a), and this code has been adapted and implemented on a transputer enhanced work-station (IBM 6150) in order to evaluate the effectiveness of parallel processing to solve integral equations.

8.2. TRANSPUTER TOPOLOGY

A very important decision is how the transputer array should be connected, this is termed the topology of the transputer array. BIM2D has been implemented on three topologies, namely, **chain**, **ring** and **cylinder**, see figure 5(b).

The results in Table 4 demonstrate the viability of transputer systems for boundary integral methodologies [78]. Of the three topologies investigated the ring topology for the array sizes and order of matrices examined is the best. This is because the extra complexity of organising the cylinder harness gives no appreciable increase in performance.

It has been recognised for some time that boundary integral methods would benefit from a weighted residual treatment. This arises for two reasons, firstly, it obviates the need to consider too closely the singularity problems of the kernels; and secondly, the matrix becomes symmetric and thus reduces storage and by implication increases the size of problem that can be solved. However the extra time involved in performing the integrations has dissuaded many workers from considering this very natural extension to the standard approach. Loosely coupled concurrent computer systems will bring new life to this very powerful computational tool. This is further assured since the costly re-construction of the fields is a perfectly parallel operation and thus the post-processing will benefit greatly from an array of processors.

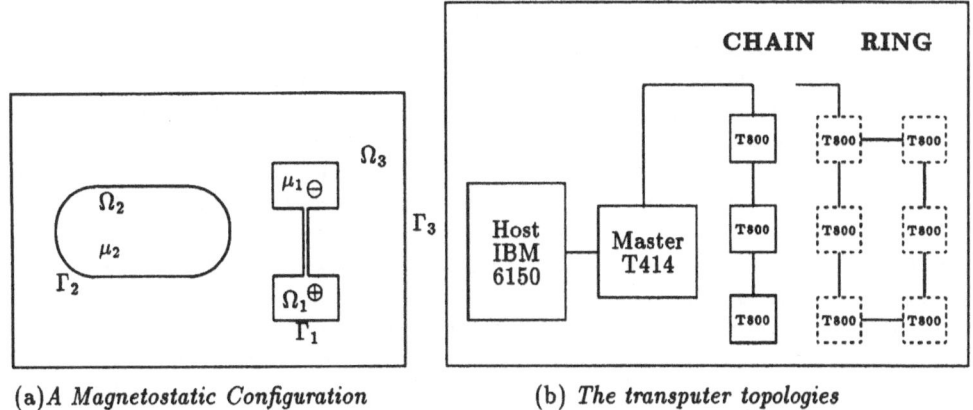

(a)*A Magnetostatic Configuration*   (b) *The transputer topologies*

Figure 5: *Boundary Element Problem and Transputer Topology*

Time in seconds for Matrix Solution					
Matrix	Number of Transputers				
Size	2	4	8	16	32
50	0.35	0.19	0.12	0.09	0.08
100	2.54	1.30	0.70	0.43	0.29
150	8.38	4.20	2.18	1.22	0.76
200	19.68	9.76	4.97	2.70	1.58
250		18.89	9.53	5.06	2.85
300		32.41	16.27	8.53	4.70
350			25.63	13.31	7.20
400			38.02	19.62	10.53
450				27.71	14.72
500				37.76	19.91
600				64.61	33.69
700					52.72
800					77.86
900					110.06

Table 4: Gaussian Elimination Times on a Ring Topology

9.   The Way Ahead?

In this lecture an outline survey has been given for the main components of a CAD system for electromagnetic design analysis. In the future it is vital that attention is given to automatic mesh generation, solution methods and error estimation. It is now possible to provide design engineers with an integrated CAD system which should enrich his array of tools for analysing electromagnetic devices. Whether this will improve the quality of design depends upon other factors but it does offer rapid methods of checking out untried ideas and a further reduction in the cost of prototype testing. As hardware becomes more powerful the range and size of problem that can be handled will increase. Of particular interest are the new developments in parallel processors and transputers which will enable solutions of large problems achievable today on large mainframes only, to be carried out on small work stations within a few years. On the other hand basic algorithm development in techniques for linear and non-linear algebra, optimisation theory and discretisation techniques will in due course extend the frontiers of computing to include complex coupled problems, involving the interaction of electromagnetic, thermal and stress tolerances. Automatic mesh generating and self adaptive schemes will progress to the point where 3D calculations can be carried out to prescribed tolerances.

Other computational techniques, not covered here are beginning to have an impact; these include the incorporation of design rules into knowledge base systems and the use of declarative languages to establish solutions to design problems not amenable to classical analysis. These considerations are becoming increasingly important as engineers begin to use the techniques of artificial intelligence to interrogate and make inferences by using the vast store of knowledge and past-experience data amassed on electrical devices of all types.

Finally it must be stressed that the results of computer simulations will only be meaningful and robust if the physical modelling is well understood and appropriate to the problem in hand; and in the end this will depend upon the judgement and inventiveness of the engineer.

10.   References

[1] J. Simkin, ed., *Compumag Oxford Proceedings*, (Chilton, Oxfordshire, UK), 1976.

[2] J. C. Sabonnadiere, ed., *Compumag Grenoble Proceedings*, (ERA 524 CNRS, Grenoble, France), 1979.

[3] L. Turner, ed., *Compumag Chicago Proceedings*, IEEE Trans Magnetics 18(2) 1982, 1981.

[4] G. Molinari, ed., *Compumag Genoa Proceedings*, IEEE Trans Magnetics 19(6) 1983, 1983.

[5] W. Lord, ed., *Compumag Fort Collins Proceedings*, IEEE Trans Magnetics 21(6) 1985, 1985.

[6] K. Richter, ed., *Compumag Graz Proceedings*, IEEE Trans Magnetics 24(1) 1988, 1987.

[7] K. Miya, ed., *Compumag Tokyo Proceedings*, IEEE Trans Magnetics 26(2) 1990, 1989.

[8] G.Molinari *et al.*, "Integrating analysis and design," in *CIM EUROPE 5th Annual Conference*, (IFS Publications, 35-39 High St., Kempston, Bedford MK42 7BT, England, UK), 1989.

[9] O.C.Zeinkiewicz, *The Finite Element Method*. Maidenhead, Berkshire, England: Mc-Graw Hill, 1977.

[10] M. J. Turner *et al.*, "Stiffness and deflection analysis of complex structures," *J Aero Sci*, vol. 23, p. 805, 1956.

[11] J. S. Hornsby, "A computer program for the solution of elliptic partial differential equations," Tech. Rep. 63-7, CERN, 1967.

[12] A. A. Winslow, "Numerical solution of the quasi-linear poisson equation in a non-uniform triangular mesh," *J Comput Phys 1*, p. 149, 1971.

[13] J. Colonias, "Calculation of 2-dimensional fields by digital display techniques," Tech. Rep. 17340, UCRL, 1967.

[14] K. Halbach and R. Holsinger, "Poisson user manual," Tech. Rep., Lawrence Berkeley Laboratory, Berkeley, CA, 1972.

[15] F. C. Trutt, *Analysis of Homopolar Inductor Alternators*. PhD thesis, University of Delaware, 1982.

[16] E. A. Erdelyi and S. V. Ahmed, "Non-linear theory of synchronous machines on load," *IEE Trans on Power Apparatus and Systems*, vol. 85, p. 792, 1966.

[17] G. Molinari and A. Viviani, "Grid and metric optimisation procedures in finite difference and finite element method," *Proc. of IEEE PES Winter Meeting, New York*, vol. A78, pp. 289–1, 1–9, 1978.

[18] W. Muller and W. Wolff, "General numerical solution of the magnetostatics equations," Tech. Rep. 49(3), AEG Telefunken, 1976.

[19] J. T. Oden, "A general theory of finite elements," *Int J Num Methh Eng*, vol. 1, no. 1, 1969.

[20] B. G. Galerkin, "Series solution of some problems of elastic equilibrium of rods and plates," *Vestn Inyh Tech*, vol. 19, p. 897, 1915.

[21] M. V. K. Chari and P. P. Silvester, "Finite element analysis of magnetically saturated dc machines," *IEEE Trans PAS*, vol. 90, p. 2362, 1971.

[22] J. Simkin and C. W. Trowbridge, "On the use of the total scalar potential in the numerical solution of field problems in electromagnets," *IJNME*, vol. 14, p. 423, 1979.

[23] J. Simkin and C. W. Trowbridge, "Three dimensional non-linear electromagnetic field computations using scalar potentials," *Proceedings of the IEE*, vol. 127, no. 6, pp. 368–374, 1980.

[24] R. F. Harrington, *Field Computation by Moment Methods*. McMillan, 1968.

[25] J.Simkin and C.W.Trowbridge, "Magnetostatic fields computed using an integral equation derived from green's theorem," in *Compumag Conference on the computation of magnetic fields*, (Chilton, Didcot, Oxon, UK), p. 5, Rutherford Appleton Laboratory, 1976.

[26] C. W. Trowbridge, *Applications of Integral Equation Methods for the Numerical Solution of Magnetostatic and Eddy Current Problems*. Chichester: Wiley, 1979.

[27] M. J. Newman, L. R. Turner, and C. W. Trowbridge, "G-FUN: An Interactive Program as an aid to Magnet Design ," *Proceedings MT4, Brookhaven*, 1972.

[28] I. M. Mayergoyz, "Mathematical models of hysteresis," *IEEE Trans Mag. Mag-22*, p. 603, 1986.

[29] D. Jiles and J. B. Thoelka, "Theory of ferromagnetic hysteresis: determination of model parameters from experimental hysteresis loops," in *Intermag Washington*, IEEE Trans on Magnetics, 1989.

[30] J.Simkin, "Modelling ferromagnetic materials for electromagnetic non destructive testing simulation," in *Third National Seminar on Non-Destructive Evaluation of Ferromagnetic Materials*, (Houston, Texas, USA), pp. 109–113, Western Atlas International, Inc., 1988.

[31] C. R. I. Emson, "Methods for the solution of open-boundary electromagnetic problems ," *IEE Proceedings, Pt A*, vol. 135, p. 152, March 1988.

[32] P.Bettes, "Infinite elements," *Int.J.Numer.Methods Eng.*, pp. 53–64, 1977.

[33] P. P. Silvester *et al.*, "Exterior finite elements for 2-dimensional field problems with open boundaries," *IEE Proceedings*, vol. 118, pp. 1743–1747, 1971.

[34] O. D. Kellogg, *Foundations of Potential Theory*. New York: Dover Publications, INC., 1954.

[35] R. Albanese and G. Rubinacci, "Solution of three dimensional eddy current problems by integral and differential methods," *IEEE Transactions on Magnetics*, vol. 24, p. 98, January 1988.

[36] Q. Xiuying and N. Guangzheng, "Electromagnetic field analysis in boundless space by finite element method," *IEEE Trans Mag-24, No 1*, 1988.

[37] E. M. Freeman and D. A. Lowther, "An open boundary technique for axisymmetric and three dimensional magnetic and electric field problems," *IEEE Trans Mag. Mag-25*, pp. 4135–4137, 1989.

[38] G. F. Carey and J. T. Oden, *Finite Elements and Computational Aspects, Vol 3*. New York: Prentice Hall, 1984.

[39] J. Bigeon *et al.*, "Finite element analysis of an electromagnetic brake," *IEEE Trans on Magnetics, Vol 19, No 6*, 1983.

[40] D. Rodger and J. Eastham, "Characteristics of a linear induction tachometer," *IEEE Trans on Magnetics, Vol 21, No 6*, 1985.

[41] A.G.A.M.Armstrong *et al.*, "Automated optimisation of magnet design using a boundary integral method," *IEEE Trans Mag. Mag-18, No 2*, 1982.

[42] K Preis and A Ziegler, " Optimal Design of Electromagnetic Devices with Evolution Strategies," in *3DMAG, International Symposium on 3-D Electromagnetic Fields Okayama, Japan, September 1989 Proceedings*, pp. 119–122, " COMPEL, Vol 9 Supp. A", 1990.

[43] K Preis and C Magele, " FEM and Evolution Strategies in the Optimal Design of Electromagnetic Devices," in *Intermag Conference, Brighton, April 1990 Proceedings*, "To be published in IEEE Trans Magnetics", 1990.

[44] M. Johnson, "Simulated Annealing," *American Journal of Mathematical and Management Sciences, Volume 8, No 3*, pp. 205–207, 1988.

[45] C.S.Biddlecombe *et al.*, "Error analysis in finite element models of electromagnetic fields," *IEEE Trans, Intermag Arizona*, 1986.

[46] P. Girdinio *et al.*, "Local error estimates for adaptive mesh refinement," *IEEE Transactions on Magnetics*, vol. 24, p. 299, January 1988.

[47] J. A. Stratton, *Electromagnetic Theory*. New York: McGraw Hill, 1941.

[48] C. S. Biddlecombe *et al.*, "Methods for Eddy Current Computation in Three-Dimensions," *IEEE Transactions on Magnetics*, vol. 18, p. 492, March 1982.

[49] A. Kameari, "Three Dimensional Eddy Current Calculation using Finite Element Method with A-V in Conductor and Ω in Vacuum," *IEEE Transactions on Magnetics*, vol. 24, January 1988.

[50] T. Morisue, "Magnetic Vector Potential and Electric Scalar Potential in Three-Dimensional Eddy Current Problem," *IEEE Transactions on Magnetics*, vol. 18, p. 531, March 1982.

[51] O. Biro, "Coulomb gauged vector potential formulation," *Proceedings of the Eddy Current Workshop, Capri*, October 1988.

[52] C F Bryant, C R I Emson and C W Trowbridge, " A Comparison of Lorentz Gauge Formulations in Eddy Current Computations," in *Compumag Conference on the Computation of Electromagnetic Fields, Tokyo, September 1989 Proceedings*, IEEE Trans Magnetics 26(2) 1990, 1990.

[53] C F Bryant, C R I Emson and C W Trowbridge, " A General Purpose 3-D Formulation for Eddy Currents using the Lorentz Gauge," in *Intermag Conference, Brighton, April 1990 Proceedings*, "To be published in IEEE Trans Magnetics", 1990.

[54] C W Trowbridge, C F Bryant and C R I Emson, " Some Recent Developments in Electromagnetic Field Computation," in *MAFELAP 1990, 7th Conference on The Mathematics of Finite Elements and Applications, April 1990,* "Academic Press, London", 1990.

[55] U. Jeske, "Eddy current calculations in 3D using the finite element method," *IEEE Transactions on Magnetics,* vol. 18, p. 426, March 1982.

[56] D. Rodger and J. F. Eastham, "A Formulation for Low Frequency Eddy Current Solutions," *IEEE Transactions on Magnetics,* vol. 19, p. 2443, November 1983.

[57] D. Rodger, "Finite element method for calculating power frequency 3 dimensional electromagnetic field distributions," *IEE Proceedings,* vol. 130, p. 233, July 1983.

[58] S. J. Polak *et al.,* "A New 3D Eddy Current Model," *IEEE Transactions on Magnetics,* vol. 19, p. 2447, November 1983.

[59] C. R. I. Emson and J. Simkin, "An Optimal Method for 3D Eddy Currents," *IEEE Transactions on Magnetics,* vol. 19, p. 2450, November 1983.

[60] A. Bossavit and J. C. Verite, "A fixed FEM-BIEM method to solve 3D eddy current probblems," *IEEE Transactions on Magnetics,* vol. 18, p. 431, March 1982.

[61] A. Bossavit and J. C. Verite, "The TRIFOU code: Solving the 3D eddy currents problem by using **H** as state variable," *IEEE Transactions on Magnetics,* vol. 19, p. 2465, November 1983.

[62] T. W. Preston and A. B. J. Reece, "Solution of 3-Dimensional eddy current problems: T-Ω method," *IEEE Transactions on Magnetics,* vol. 18, p. 486, March 1982.

[63] R. Albanese, R. Martone, G. Miano, and G. Rubinacci, "A **T** formulation for 3D finite element eddy current computation," *IEEE Transactions on Magnetics,* vol. 21, p. 2299, November 1985.

[64] T. Nakata, N. Takahashi, K. Fujiwara, and Y. Okada, "Improvements of the T-Ω method for 3-D eddy current analysis," *IEEE Transactions on Magnetics,* vol. 24, p. 94, January 1988.

[65] J. Wang and B. Tong, "Calculation of 3-D eddy current problems using a modified T-Ω method," *IEEE Transactions on Magnetics,* vol. 24, p. 114, January 1988.

[66] C. J. Carpenter, "Comparison of alternative formulations of 3-D magnetic field and eddy current problems at power frequencies," *Proceedings IEE,* vol. 124, no. 11, p. 1026, 1977.

[67] R. Albanese and G. Rubinacci, "Integral formulation for 3D eddy current computation using edge elements," *IEE Proceedings,* vol. 135, p. 457, September 1988.

[68] C. W. Trowbridge, "Electromagnetic computing: the way ahead," *IEEE Transactions on Magnetics,* vol. 24, p. 13, January 1988.

[69] J. A. M. Davidson and M. J. Balchin, "3-D field calculations by network methods," *IEEE Transactions on Magnetics*, vol. 19, November 1983.

[70] D. W. Weissenburger and U. R. Christenson, "A network mesh method to calculate eddy currents on conducting surfaces," *IEEE Transactions on Magnetics*, vol. 18, March 1982.

[71] L. R. Turner and R. J. Lari, "Applications and further developments of the eddy current program EDDYNET," *IEEE Transactions on Magnetics*, vol. 18, 1982.

[72] L. R. Turner *et al.*, "Papers on bench-mark problems for the validation of eddy current codes," *COMPEL*, vol. 7, March/June 1982.

[73] R. W. Hockney and C. R. Jesshope, *Parallel Computers (Architectures, Programming and Algorithms)*. Bristol, UK: Adam Hilger Ltd., 1981.

[74] *TOSCA, GFUN, CARMEN, ELEKTRA, BIM2D, OPERA, and PE2D User Manuals*. Vector Fields Ltd., 24 Bankside, Kidlington, Oxford OX5 1JE, 1988.

[75] N. J. Diserens and A.R.Mayhook, "Experience in the use of Vector Processors for 3-D Static Analysis," in *Compumag Conference on the Computation of Electromagnetic Fields, Tokyo, September 1989 Proceedings*, "To be published in IEEE Trans Magnetics", 1990.

[76] C.P.Riley, "Solution of EM Fields on Vector Processors," *VECTOR Electromagnetics Newsletter*, vol. 4, no. 1, 1988.

[77] C.F.Bryant and C.W.Trowbridge, "Specification of the APPEAL Library," Tech. Rep., Vector Fields Ltd., 24 Bankside, Kidlington, Oxford OX5 1JE, 1988. Esprit(1051), ACCORD/WP4/DEL/VFL/003/26.07.88/CFB/CWT/.

[78] C. F. Bryant *et al.*, "Implementing a boundary integral method on a transputer system," in *Compumag Conference on the Computation of Electromagnetic Fields, Tokyo, September 1989 Proceedings*, "To be published in IEEE Trans Magnetics", 1990.

[79] "Transputer technical information,". INMOS Ltd., 1985, PO BOX 424, Bristol, BS99 7DD, UK.

NUMERICAL FOUNDATIONS, INTEGRAL METHODS AND APPLICATIONS

K. R. RICHTER and W. M. RUCKER
Institute for Fundamentals and Theory in Electrical Engineering
Graz University of Technology
Kopernikusgasse 24
A-8010 Graz, Austria

ABSTRACT. In this paper the mathematical foundations and the application of various integral formulations based on vector and scalar potentials for the numerical treatment of magnetostatic and eddy current problems with the boundary element method (BEM) are presented. Starting from Maxwell's equations the basic integral equations are derived by applying Green's theorem for scalar and vector variables and the definition of the appropriate Green's functions. In order to apply the BEM the discretization of the boundaries into boundary elements is discussed. This discretization transforms the boundary integral equations into a system of algebraic equations which can be solved by various methods. The boundary element formulations and numerical results of examples for three-dimensional magnetostatic field problems and eddy current problems are presented. It can be seen that there exists a wide range of applications of integral methods and which are alternatives to differential equation methods, e.g. the finite element method.

1. INTRODUCTION

An electromagnetic field problem in general is defined by a partial differential equation and appropriate boundary and initial conditions. There exist only a few simple problems with analytical solutions. By defining a Green's function which satisfies the differential equation and the boundary conditions for a point source, the solution of the field problem can be expressed as an integral over the known source function multiplied by Green's function. With this method the differential equation can be transformed into an integral equation which sometimes can be solved easier.

Since during the last thirty years the computer power increased extensively the numerical methods became more and more attractive. Especially the finite element method (FEM) has been established as a very powerful tool for the design of electromagnetic devices.

In the early sixties R. Harrington [1] developed the so-called method of moments for the numerical solution of integral equations. With this method the integral equations are transferred into algebraic equations by introducing test and weighting functions for the unknown field.

Some years later C. Brebbia [2] created the so-called boundary element

29

Y. R. Crutzen et al. (eds.), Industrial Application of Electromagnetic Computer Codes, 29–50.
© 1990 *ECSC, EEC, EAEC, Brussels and Luxembourg.*

method (BEM) where the integral equations are solved by discretizing the boundary into finite elements, the boundary elements. From the modellisation of the field problem the BEM is very similar to the FEM, however the resulting equations systems are strongly different. The BEM is characterized by low modelling effort and small but fully populated, non-symmetric coefficient matrices, while the FEM results in a large equations system with sparse coefficient matrices. Today it seems that both methods are complementary partners and not opponents.

The aim of this paper is to collect the various integral formulations for the calculation of two-dimensional (2d) and three-dimensional (3d) magnetostatic and eddy current problems and to show how to solve them by BEM.

In the first part the governing equations for magnetostatic field problems using vector and scalar potential formulations are derived from Maxwell's equations. Emphasis is put on the interface conditions between regions of different material properties. For 3d eddy current problems with harmonic time variation a coupled magnetic vector and electric scalar potential formulation is presented. The section ends with the basic equations for transient 2d problems.

In the second chapter the expressions for a boundary element formulation using second order isoparametric elements are derived.

In the last part the application of the BEM is shown for some examples and numerical results are presented.

As conclusion the relationships developed in this paper are not restricted to magnetostatic and eddy current problems, they can be extended to electrostatic and electrokinetic problems as well as to electromagnetic wave problems (e.g. scattering problems).

2. MATHEMATICAL FOUNDATIONS

2.1. MAGNETOSTATIC

2.1.1. *Magnetic vector potential formulation.* Starting from Maxwell's equations for a steady-state current flow

$$\nabla \times \mathbf{H} = \mathbf{J} \qquad (1)$$

$$\nabla . \mathbf{B} = 0 \qquad (2)$$

the magnetic vector potential $\mathbf{A}$ is defined by

$$\mathbf{B} = \nabla \times \mathbf{A} \quad . \qquad (3)$$

With the constitutive law

$$\mathbf{B} = \mu \mathbf{H} \qquad (4)$$

for a constant magnetic permeability μ the differential equation

$$\nabla \times (\nabla \times \mathbf{A}) = \mu \mathbf{J} \qquad (5)$$

can be derived.

The magnetic vector potential is not uniquely defined by eq.(3).
Introducing the Coulomb gauge

$$\nabla.\mathbf{A} = 0 \tag{6}$$

eq.(5) can be written as

$$\nabla^2\mathbf{A} = -\mu\mathbf{J} . \tag{7}$$

The free space solution of this equation is given by

$$\mathbf{A(r)} = \int\int\int_{-\infty}^{\infty} \mathbf{J(r)} \, G(\mathbf{r,r'}) \, d\Omega \tag{8}$$

where the free space Green's function

$$G(\mathbf{r,r'}) = \frac{1}{4\pi|\mathbf{r}-\mathbf{r'}|} \tag{9}$$

satisfies the equation

$$\nabla^2 G(\mathbf{r,r'}) = -\delta(\mathbf{r}-\mathbf{r'}) \tag{10}$$

describing the static field of a point source,

$$\delta(\mathbf{r}-\mathbf{r'}) = \begin{cases} 0 & \text{for } \mathbf{r} \neq \mathbf{r'} \\ \infty & \text{for } \mathbf{r} = \mathbf{r'} \end{cases} \tag{11}$$

represents the Delta function.
In the case where a finite volume Ω bounded by the surface Γ is considered (Fig.1), the field quantities on the boundary must be taken into account.

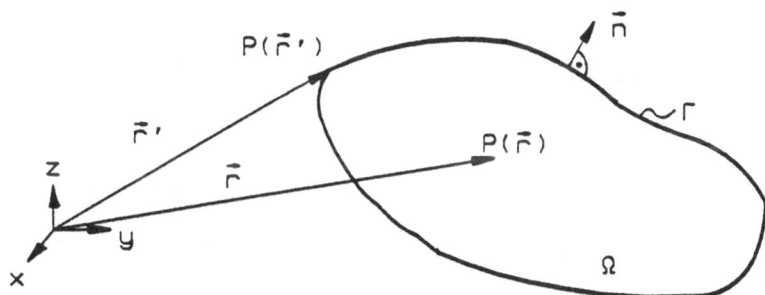

Figure 1. Geometry of a closed problem.

Therefore, an integral representation of the magnetic vector potential $\mathbf{A}$ is obtained by using the vector equivalent of Green's theorem

$$\int\int\int_{\Omega} [\mathbf{A}.\nabla \times \nabla \times \mathbf{G} - \mathbf{G}.\nabla \times \nabla \times \mathbf{A}] \, d\Omega = \oiint_{\Gamma} [\mathbf{G} \times \nabla \times \mathbf{A} - \mathbf{A} \times \nabla \times \mathbf{G}].\mathbf{n} \, d\Gamma \tag{12}$$

Using a vector Green's function

$$G(r,r') = G(r,r') \, p \, , \tag{13}$$

where p is an arbitrary constant vector and $G(r,r')$ from eq.(9) which satisfies

$$\nabla^2 G(r,r') = -\delta(r - r') \, p \tag{14}$$

or

$$\nabla \times \nabla \times G(r,r') = \nabla\nabla.\, G(r,r') + \delta(r - r') \, p \tag{15}$$

the volume integral in eq.(12) becomes

$$\iiint_\Omega [A.\nabla \times \nabla \times G - G.\nabla \times \nabla \times A]\,d\Omega = A.p + \iiint_\Omega A.\nabla(\nabla.G)\,d\Omega - \mu\iiint_\Omega J.G\,d\Omega \tag{16}$$

The first integral term in eq.(16) can be integrated by part

$$\iiint_\Omega A.\nabla(\nabla.G)\,d\Omega = \oiint_\Gamma \nabla.G(A.n)\,d\Gamma - \iiint_\Omega (\nabla.A)\,(\nabla.G)\,d\Omega \tag{17}$$

and is transformed into a surface integral term since the Coulomb gauge is active and therefore the volume term of the right side of eq.(17) vanishes.

The second integral term in eq.(16) can be interpreted as an impressed magnetic vector potential A^i due to the current density within the region Ω:

$$A^i = \mu\iiint_\Omega J\,G\,d\Omega. \tag{18}$$

Equation (16) becomes

$$\iiint_\Omega [A.\nabla \times \nabla \times G - G.\nabla \times \nabla \times A]\,d\Omega = p.\left\{ A + \oiint_\Gamma (A.n)\,\nabla G\,d\Gamma - A^i \right\} . \tag{19}$$

Using the vector identities

$$(G \times \nabla \times A).n = G(n \times \nabla \times A).p \tag{20}$$

and

$$(A \times \nabla \times G).n = -(n \times \nabla \times G).A = (\nabla G \times n \times A).p \tag{21}$$

the surface integral of eq.(12) can be written as

$$\oiint_\Gamma [G \times \nabla \times A - A \times \nabla \times G].n\,d\Gamma = p.\oiint_\Gamma [(n \times \nabla \times A)G + (n \times A) \times \nabla G]\,d\Gamma. \tag{22}$$

Combining the equations (19) and (22) and dividing by p the magnetic vector potential for points within the domain Ω can be expressed as

$$A(r) = -\oiint_\Gamma [(n \times \nabla \times A)G + (n \times A) \times \nabla G + (n.A)\nabla G]\,d\Gamma + A^i(r) . \tag{23}$$

The integral equation for A is obtained by considering the field points $P(r)$ on the surface Γ. In this the kernels of the surface integral become singular because $\mathbf{r}$ and $\mathbf{r}'$ coincide. The surface integral of eq.(23) exists in a Cauchy sense and the contribution of the singularity can be calculated as

$$I_\varepsilon = \lim_{\varepsilon \to 0} \iint_{\Gamma_\varepsilon} [(\mathbf{n} \times \nabla \times \mathbf{A})G + (\mathbf{n} \times \mathbf{A}) \times \nabla G + (\mathbf{n}.\mathbf{A})\nabla G] d\Gamma = (-1 + \frac{\Theta}{4\pi})\mathbf{A}(\mathbf{r}) \tag{24}$$

where Θ is the spatial angle which has for a smooth boundary the value $\Theta = 2\pi$, so that $I_\varepsilon = -1/2$.

Therefore the integral equation can be written as

$$c(\mathbf{r})\mathbf{A}(\mathbf{r}) + \oiint_{\Gamma} [(\mathbf{n}' \times \mathbf{A}(\mathbf{r}')) \times \nabla G(\mathbf{r},\mathbf{r}') + (\mathbf{n}.\mathbf{A}(\mathbf{r}')) \nabla G(\mathbf{r},\mathbf{r}')] d\Gamma =$$

$$= -\oiint_{\Gamma} (\mathbf{n} \times \nabla \times \mathbf{A}(\mathbf{r}')) G(\mathbf{r},\mathbf{r}') d\Gamma + \mathbf{A}^i(\mathbf{r}) \tag{25}$$

with

$$c(\mathbf{r}) = \frac{\Theta(\mathbf{r})}{4\pi} . \tag{26}$$

At the interface between two regions $\Omega_1 (\mu_1)$ and $\Omega_2 (\mu_2)$ the continuity of the normal component of the flux density $\mathbf{B}$ and the tangential component of the magnetic excitation $\mathbf{H}$ results in

$$\mathbf{A}_1 = \mathbf{A}_2 \tag{27}$$

and

$$\frac{1}{\mu_1}\mathbf{n} \times \nabla \times \mathbf{A}_1 = \frac{1}{\mu_2}\mathbf{n} \times \nabla \times \mathbf{A}_2 . \tag{28}$$

If a plane problem $(\partial/\partial z = 0)$ is considered where the current density $\mathbf{J}$ is perpendicular to the xy-plane:

$$\mathbf{J} = J(x,y)\,\mathbf{e}_z \tag{29}$$

the magnetic vector potential $\mathbf{A}$ has only one component

$$\mathbf{A} = A(x,y)\,\mathbf{e}_z \tag{30}$$

and therefore the vector boundary integral equation (25) is reduced to a simple scalar integral equation:

$$c(\mathbf{r})A(\mathbf{r}) + \oint_{\Gamma} A(\mathbf{r}')\frac{\partial G(\mathbf{r},\mathbf{r}')}{\partial n'} d\Gamma = \oint_{\Gamma} \frac{\partial A(\mathbf{r}')}{\partial n'} G(\mathbf{r},\mathbf{r}') d\Gamma + A^i(\mathbf{r}) \tag{31}$$

where $\partial/\partial n$ means $\mathbf{n}.\nabla$.

In eq.(31) the boundary Γ encloses the domain Ω which represents the cross-section of a cylindrical problem.

The contribution of the singularity leads to

$$c(\mathbf{r}) = \frac{\Theta(\mathbf{r})}{2\pi} = - \oint_{\Gamma} \frac{\partial G(\mathbf{r},\mathbf{r}')}{\partial n'} d\Gamma . \tag{32}$$

The Green's function for the plane case has the form

$$G(r,r') = \frac{1}{2\pi} \ln \frac{1}{|r-r'|} \quad . \tag{33}$$

The conditions on an interface between regions with different permeabilities are

$$A_1 = A_2 \quad , \tag{34}$$

$$\frac{1}{\mu_1} \frac{\partial A_1}{\partial n} = \frac{1}{\mu_2} \frac{\partial A_2}{\partial n} \quad . \tag{35}$$

For axisymmetric problems $(\partial/\partial\varphi = 0)$ which are characterized by

$$J = J(R,z)\, e_\varphi \tag{36}$$

and

$$A = A(R,z)\, e_\varphi \tag{37}$$

the boundary integral equation has a similar form as eq.(31):

$$c(r)A(r) + \oint_\Gamma A(r') \frac{\partial G(r,r')}{\partial n'} d\Gamma = \oint_\Gamma \frac{1}{R'} \frac{\partial(R'A(r'))}{\partial n'} G(r,r') d\Gamma + A^i(r) \quad . \tag{38}$$

Again the boundary Γ is the closed contour of the cross-section in the symmetry plane (R,z) shown in Fig.2.

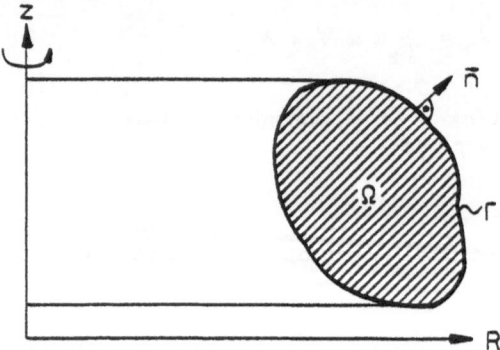

Figure 2. Cross-section of an axisymmetric problem.

The corresponding free space Green's function for this axisymmetric vector field is

$$G(r,r') = \frac{1}{\pi k} \sqrt{\frac{R'}{R}} \left[(1 - \frac{k^2}{2}) K(k) - E(k) \right] \tag{39}$$

where

$$k = \sqrt{\frac{4RR'}{(z-z')^2 + (R+R')^2}} \tag{40}$$

and K(k), E(k) are the complete elliptic integrals of the first and second kind.

The interface conditions are

$$A_1 = A_2 \quad , \tag{41}$$

$$\frac{1}{\mu_1} \frac{\partial (RA_1)}{\partial n} = \frac{1}{\mu_2} \frac{\partial (RA_2)}{\partial n} \quad . \tag{42}$$

2.1.2. *Scalar potential formulation.* For the calculation of a three-dimensional magnetostatic field it is much more economic to use a magnetic scalar potential formulation than a magnetic vector potential formulation (eq.(25)) [4]. However, the magnetic field can be expressed as a gradient of a scalar potential only in current-free regions. Therefore, the total magnetic field must be splitted up into two components

$$H = H_M + H_S \quad , \tag{43}$$

where

$$H_M = -\nabla \Phi \tag{44}$$

represents the curl-free induced magnetic field, which can be expressed as a gradient of the so-called reduced scalar potential Φ.

The magnetic field H_S is a solenoidal field due to current sources and may be calculated by Biot-Savart's law:

$$H_S = \int_{\Omega_s} \frac{I \times (r - r')}{4\pi |r - r'|^3} d\Omega \quad . \tag{45}$$

For a constant permeability μ and according to equations (2) and (4) the reduced scalar potential satisfies the Laplace equation

$$\nabla^2 \Phi = 0 \quad . \tag{46}$$

Taking Green's function of eq.(9) and applying Green's theorem

$$\iiint_{\Omega} [\Phi \nabla^2 G - G \nabla^2 \Phi] d\Omega = \oiint_{\Gamma} [\Phi \frac{\partial G}{\partial n} - G \frac{\partial \Phi}{\partial n}] d\Gamma \tag{47}$$

for the one component of the magnetic vector potential the same boundary integral equation as eq.(31) can be derived:

$$c(r) \Phi(r) + \oiint_{\Gamma} \Phi(r') \frac{\partial G(r,r')}{\partial n'} d\Gamma = \oiint_{\Gamma} \frac{\partial \Phi(r')}{\partial n'} G(r,r') d\Gamma \quad . \tag{48}$$

At the interface between regions with different permeabilities the conditions

$$\Phi_1 = \Phi_2 \quad , \tag{49}$$

$$\mu_1 \frac{\partial \Phi}{\partial n} - \mu_1 H_S \cdot n = \mu_2 \frac{\partial \Phi}{\partial n} - \mu_2 H_S \cdot n \tag{50}$$

must be satisfied.

For the computation of a magnetostatic field problem the two integral equations (eq.(48)) formulated for the two different regions $\Omega_1(\mu_1)$ and $\Omega_2(\mu_2)$ must be coupled at the interface $\Gamma_{1,2}$ with the conditions (49) and (50). From this follows an source term:

$$S_\Phi(r) = \frac{\mu_1 - \mu_2}{\mu_2} \oint\limits_\Gamma H_S(r') . n \; G(r,r') \, d\Gamma \; . \tag{51}$$

The practical application of this reduced scalar potential formulation has shown, that within regions with a high permeability ($\mu_r > 1000$) the calculation of the total magnetic field leads to very inaccurate results caused by cancellation errors in applying eq.(43) [4].

These problems can be avoided by defining a so-called total scalar potential ψ. In this case the total magnetic field H can be expressed as

$$H = -\nabla \psi \; . \tag{52}$$

This formulation is allowed only in regions without currents. For current carrying regions the reduced scalar potential formulation must be used. Since usually these regions have a low permeability no numerical difficulties will arise.

The total scalar potential also satisfies the Laplace equation

$$\nabla^2 \psi = 0 \; . \tag{53}$$

Therefore the same boundary integral equation as eq.(48) can be derived for the total scalar potential.

At the interface between a ψ-region (e.g. iron core) and a Φ-region (e.g. air domain) the continuity conditions for the magnetic field lead to

$$\psi(r) = \Phi(r) - \int\limits_{P_0}^{P(r)} H_S(r') . t' \, dt \; , \tag{54}$$

$$\mu_1 \frac{\partial \psi}{\partial n} = \mu_2 \frac{\partial \psi}{\partial n} - \mu_2 H_S . n \; . \tag{55}$$

It can be seen that in eq.(54) a line integral over the tangential component of the source field H_S exists. The arbitrary integration path is situated in the interface between the two regions. Care must be taken, that no current carrying region is enclosed by the integration path.

For the numerical calculations the coupling of the two different regions with their different formulations lead to the following source term:

$$S_\psi(r) = -c(r) \int\limits_{P_0}^{P(r)} H_S(r') . t' \, dt + \oint\limits_\Gamma [\int\limits_{P_0}^{P(r')} H_S(r'') . t'' \, dt] \frac{\partial G(r,r')}{\partial n'} d\Gamma -$$

$$- \oint\limits_\Gamma H_S(r') . n \; G(r,r') \, d\Gamma \; . \tag{56}$$

If the anisotropic behaviour of a magnetic material is considered the magnetic permeability will become a tensor $[\mu]$, so that

$$\mathbf{B} = [\mu]\mathbf{H} . \tag{57}$$

Therefore, the differential equation e.g. for the total scalar potential is

$$\nabla [\mu] \nabla \psi = 0 . \tag{58}$$

The boundary integral equations remain unchanged, except that the Green's function has to satisfy

$$\nabla [\mu] \nabla G(\mathbf{r},\mathbf{r}') = -\delta(\mathbf{r} - \mathbf{r}') . \tag{59}$$

For example for an orthotropic medium with

$$[\mu] = \begin{bmatrix} \mu_x & 0 & 0 \\ 0 & \mu_y & 0 \\ 0 & 0 & \mu_z \end{bmatrix} \tag{60}$$

the Green's function for a 3d problem is

$$G(\mathbf{r},\mathbf{r}') = \frac{1}{4\pi\sqrt{\mu_x\mu_y\mu_z}\ \sqrt{\dfrac{(x-x')^2}{\mu_x} + \dfrac{(y-y')^2}{\mu_y} + \dfrac{(z-z')^2}{\mu_z}}} \tag{61}$$

and for a plane 2d problem

$$G(\mathbf{r},\mathbf{r}') = \frac{1}{2\pi\sqrt{\mu_x\mu_y}} \ \ln \frac{1}{\sqrt{\dfrac{(x-x')^2}{\mu_x} + \dfrac{(y-y')^2}{\mu_y}}} \tag{62}$$

2.2. EDDY CURRENT PROBLEMS

2.2.1. *Three-dimensional problems.* For a time-harmonic, quasistationary field problem from Maxwell's equations

$$\nabla \times \mathbf{H} = \mathbf{J} , \tag{63}$$

$$\nabla \times \mathbf{E} = -j\omega\mathbf{B} \tag{64}$$

completed by the constitutive laws

$$\mathbf{B} = \mu\mathbf{H} \tag{65}$$

$$\mathbf{J} = \gamma\mathbf{E} \tag{66}$$

the following differential equation for the magnetic vector potential $\mathbf{A}$, defined in eq.(3), and for the electric scalar potential V is obtained:

$$\nabla \times \nabla \times \mathbf{A} + j\omega\mu\gamma\mathbf{A} - \mu\gamma\nabla V = \mathbf{0} .\qquad(67)$$

Introducing the Lorentz gauge

$$\nabla.\mathbf{A} + \mu\gamma V = 0\qquad(68)$$

a vector Helmholtz equation for the magnetic vector potential $\mathbf{A}$

$$\nabla^2\mathbf{A} - j\omega\mu\gamma\mathbf{A} = \mathbf{0}\qquad(69)$$

and with

$$\nabla.\mathbf{J} = 0\qquad(70)$$

a scalar Helmholtz equation for the electric scalar potential V

$$\nabla^2 V - j\omega\mu\gamma V = 0\qquad(71)$$

can be derived.

By defining a free space vector Green's function

$$G(\mathbf{r},\mathbf{r'}) = \frac{e^{-jk|\mathbf{r}-\mathbf{r'}|}}{4\pi|\mathbf{r}-\mathbf{r'}|}\,\mathbf{p}\qquad(72)$$

which satisfies

$$\nabla^2 G(\mathbf{r},\mathbf{r'}) + k^2\,G(\mathbf{r},\mathbf{r'}) = -\delta(\mathbf{r}-\mathbf{r'})\,\mathbf{p} ,\qquad(73)$$

where $\qquad k = \sqrt{-j\omega\mu\gamma}\qquad(74)$

is the complex diffusion constant, a boundary integral representation for the magnetic vector potential can be derived using again the vector equivalent of Green's theorem eq.(12).

With equations (67) and (73) the volume integral in eq.(12) can be written as

$$\iiint\limits_{\Omega}[\mathbf{A}.\nabla\times\nabla\times G - G.\nabla\times\nabla\times\mathbf{A}]\,d\Omega = \mathbf{A}.\mathbf{p} + \iiint\limits_{\Omega}\mathbf{A}.\nabla(\nabla.G)\,d\Omega -$$
$$-\mu\gamma\iiint\limits_{\Omega}G.\nabla V\,d\Omega .\qquad(75)$$

The first volume integral term in eq.(75) can be transformed by using Gauss' theorem into

$$\iiint\limits_{\Omega}\mathbf{A}.\nabla(\nabla.G)\,d\Omega = \oiint\limits_{\Gamma}\nabla.G(\mathbf{A}.\mathbf{n})\,d\Gamma - \iiint\limits_{\Omega}(\nabla.\mathbf{A})(\nabla.G)\,d\Omega .\qquad(76)$$

The second volume integral on the right-hand side in eq.(75) and the volume integral on the right-hand side of eq.(76) can be combined and transformed by using the Lorentz gauge and Gauss' theorem into a surface integral:

$$\iiint\limits_{\Omega}[-\nabla.\mathbf{A}\nabla.G + \mu\gamma G.\nabla V]\,d\Omega = \mu\gamma\mathbf{p}.\iiint\limits_{\Omega}\nabla(VG)\,d\Omega = \mu\gamma\mathbf{p}.\oiint\limits_{\Gamma}VG\mathbf{n}\,d\Gamma .\qquad(77)$$

The evaluation of the right-hand side of eq.(12) is shown in eq.(22).

Combining the equations (75), (76), (77) and (22) and dividing the result by the arbitrary constant vector **p** an integral representation for the magnetic vector potential is obtained:

$$\mathbf{A(r)} = - \oiint_{\Gamma} [(\mathbf{n} \times \nabla \times \mathbf{A}) G + (\mathbf{n} \times \mathbf{A}) \times \nabla G + (\mathbf{n} \cdot \mathbf{A}) \nabla G] d\Gamma - \mu\gamma \oiint_{\Gamma} V G \mathbf{n} d\Gamma. \quad (78)$$

For points on the surface Γ and considering the singular kernels the integral equation

$$c(\mathbf{r}) \mathbf{A(r)} + \oiint_{\Gamma} [(\mathbf{n}' \times \mathbf{A(r')}) \times \nabla G(\mathbf{r,r'}) + (\mathbf{n} \cdot \mathbf{A(r')}) \nabla G(\mathbf{r,r'})] d\Gamma =$$

$$= - \oiint_{\Gamma} (\mathbf{n} \times \nabla \times \mathbf{A(r')}) G(\mathbf{r,r'}) d\Gamma - \mu\gamma \oiint_{\Gamma} V G \mathbf{n} d\Gamma \quad (79)$$

can be derived.

The electric scalar potential V obeys the well-known scalar integral equation

$$c(\mathbf{r}) V(\mathbf{r}) + \oiint_{\Gamma} V(\mathbf{r'}) \frac{\partial G(\mathbf{r,r'})}{\partial n'} d\Gamma = \oiint_{\Gamma} \frac{\partial V(\mathbf{r'})}{\partial n'} G(\mathbf{r,r'}) d\Gamma \quad (80)$$

with

$$G(\mathbf{r,r'}) = \frac{e^{-jk|\mathbf{r-r'}|}}{4\pi |\mathbf{r-r'}|} . \quad (81)$$

From these two coupled boundary integral equations the eddy current density can be calculated, whereby

$$\mathbf{J} = -j\omega\gamma \mathbf{A} - \gamma \nabla V . \quad (82)$$

At the interface between a conducting and a non-conducting region the normal component of the current density vanishes and therefore the relationship

$$\mathbf{A} \cdot \mathbf{n} = -\frac{1}{j\omega} \frac{\partial V}{\partial n} \quad (83)$$

is valid. Beyond it the interface conditions (27) and (28) must be considered.

In the non-conducting, eddy current-free, regions the Helmholtz equation (69) is reduced to the Laplace equation and the magnetic vector potential can be described by eq.(25) for the magnetostatic case. Outside of the eddy current region the electric scalar potential is zero.

The degrees of freedom can be reduced by using other formulations (e.g. boundary integral method of minimum order [5]) instead of this **A**, V-formulation.

2.2.2. *Plane problems.* In case of a two-dimensional plane eddy current problem $(\partial / \partial z = 0)$

$$\mathbf{A} = A(x,y) \mathbf{e}_z \quad (84)$$

the Coulomb gauge

$$\nabla \cdot \mathbf{A} = 0 \quad (85)$$

and
$$\nabla V = 0 \tag{86}$$

are automatically fulfilled. Therefore, no electric scalar potential is used within the regions.

The single component of the magnetic vector potential satisfies the scalar Helmholtz equation

$$\nabla^2 A(\mathbf{r}) + k^2 A(\mathbf{r}) = 0 . \tag{87}$$

Applying Green's theorem for scalar variables the boundary integral equation

$$c(\mathbf{r}) A(\mathbf{r}) + \oint_\Gamma A(\mathbf{r}') \frac{\partial G(\mathbf{r},\mathbf{r}')}{\partial n'} d\Gamma = \oint_\Gamma \frac{\partial A(\mathbf{r}')}{\partial n'} G(\mathbf{r},\mathbf{r}') d\Gamma \tag{88}$$

can be derived. In this case the Green's function

$$G(\mathbf{r},\mathbf{r}') = \frac{1}{4j} H_0^{(2)}(k|\mathbf{r}-\mathbf{r}'|) \tag{89}$$

is a solution of

$$\nabla^2 G(\mathbf{r},\mathbf{r}') + k^2 G(\mathbf{r},\mathbf{r}') = -\delta(\mathbf{r}-\mathbf{r}') , \tag{90}$$

wherby $H_0^{(2)}(x)$ represents the Hankel function of zero order.

Using the complex Poynting vector defined as

$$T = \frac{1}{2} E \times H^* \tag{91}$$

with

$$E = -j\omega A e_z , \tag{92}$$

$$H^* = \frac{1}{\mu} \nabla A^* \times e_z \tag{93}$$

for the plane case, the average of the power losses P_0 and the stored magnetic energy W_m

$$P_0 = -\frac{\omega}{2\mu} \text{Im}\left\{ \oint A \frac{\partial A^*}{\partial n} d\Gamma \right\} , \tag{94}$$

$$W_m = \frac{1}{2\mu} \text{Re} \left\{ \oint A \frac{\partial A^*}{\partial n} d\Gamma \right\} \tag{95}$$

can be calculated from the boundary values A and $\partial A/\partial n$, which are obtained by solving eq.(88).

2.2.3. *Transient plane problems.* In the general case of a transient eddy current problem the differential equation for the time-dependent magnetic vector potential is the diffusion equation

$$\nabla^2 A(x,y,t) - \mu\gamma \frac{\partial A(x,y,t)}{\partial t} = 0 . \tag{96}$$

There exist several methods to obtain a boundary integral solution of eq.(96). The simplest method is the time stepping method, which is a finite difference procedure in the time domain

$$\frac{\partial A(x,y,t)}{\partial t} \sim \frac{A(x,y,t) - A(x,y,t-\Delta t)}{\Delta t} \tag{97}$$

and eq.(96) becomes

$$\nabla^2 A(x,y,t) - \frac{\mu\gamma}{\Delta t} A(x,y,t) = - \frac{\mu\gamma}{\Delta t} A(x,y,t-\Delta t) \quad . \tag{98}$$

The Green's function for this problem is time-independent and satisfies

$$\nabla^2 G(\mathbf{r,r'}) - \alpha^2 G(\mathbf{r,r'}) = -\delta(\mathbf{r-r'}) \tag{99}$$

with

$$G(\mathbf{r,r'}) = \frac{1}{2\pi} K_0(\alpha|\mathbf{r-r'}|) \tag{100}$$

where $K_0(x)$ represents the modified Bessel function of zero order with the real diffusion constant α

$$\alpha = \sqrt{\frac{\mu\gamma}{\Delta t}} \quad . \tag{101}$$

The boundary integral equation which can be derived from the equations (98) and (99) using Green's theorem has the form

$$c(\mathbf{r})A(\mathbf{r},t) + \oint_\Gamma A(\mathbf{r'},t) \frac{\partial G(\mathbf{r,r'})}{\partial n'} d\Gamma = \oint_\Gamma \frac{\partial A(\mathbf{r'},t)}{\partial n'} G(\mathbf{r,r'}) d\Gamma + \frac{\mu\gamma}{\Delta t} \int\!\!\int_\Omega A(\mathbf{r'},t-\Delta t) G(\mathbf{r,r'}) d\Omega \tag{102}$$

In eq.(102) there appears an integral over the domain Ω with the magnetic vector potential of the time $t-\Delta t$. For every time step the integral equation (102) must be solved.

Numerical results with higher accuracy are obtained by using a time-dependent Green's function [6]

$$G(\mathbf{r},t,\mathbf{r'},t') = \frac{1}{4\pi(t-t')} e^{-\frac{\alpha^2|\mathbf{r-r'}|^2}{4(t-t')}} \tag{103}$$

which satisfies

$$\nabla^2 G(\mathbf{r,r'}) - \alpha^2 G(\mathbf{r,r'}) = -\delta(\mathbf{r-r'})\delta(t-t') \tag{104}$$

for two-dimensional plane problems.

Applying Green's theorem completed by the time integration

$$\int_t \int\!\!\int_\Omega [A\nabla^2 G - G\nabla^2 A] d\Omega\, dt = \int_t \oint_\Gamma [A \frac{\partial G}{\partial n} - G \frac{\partial A}{\partial n}] d\Gamma\, dt \tag{105}$$

and with the equations (96) and (104) the left-hand side of eq.(105) can be written as

$$\int_{t'=0}^t \int\!\!\int_\Omega [A\nabla^2 G - G\nabla^2 A] d\Omega\, dt' = -A(\mathbf{r},t) - \alpha^2 \int_{t'=0}^t \int\!\!\int_\Omega [A \frac{\partial G}{\partial t'} + G \frac{\partial A}{\partial t'}] d\Omega\, dt' \quad . \tag{106}$$

After partial integration

$$\int_{t'=0}^{t} A \frac{\partial G}{\partial t'} dt' = A G \Big|_{t'=0}^{t'=t} - \int_{t'=0}^{t} \frac{\partial A}{\partial t'} G dt' \tag{107}$$

and with

$$G(r,t,r',t') \Big|_{\substack{t=t' \\ r \neq r'}} = 0 \tag{108}$$

the boundary integral equation for time-dependent variables is obtained:

$$c(r) A(r,t) + \int_0^t \oint_\Gamma A(r',t') \frac{\partial G(r,t,r',t')}{\partial n'} d\Gamma dt' = \int_0^t \oint_\Gamma \frac{\partial A(r',t')}{\partial n'} G(r,t,r',t') d\Gamma dt' +$$

$$+ \alpha^2 \int\int_\Omega A(r',0) G(r,t,r',0) d\Omega . \tag{109}$$

Additional to the well-known boundary integral terms again a domain integral appears, which takes the initial conditions of the magnetic vector potential into account.

3. BOUNDARY ELEMENT FORMULATION

The numerical treatment of the various boundary integral equations shown in the chapter before can be performed by applying the boundary element method (BEM). This means, that the closed surface Γ of the considered region Ω has to be discretized into finite elements, the so-called boundary elements. In Fig.3 the discretization of the boundary of a two-dimensional plane field problem is shown. In this example geometrically linear elements with a single node in the center of the elements are used.

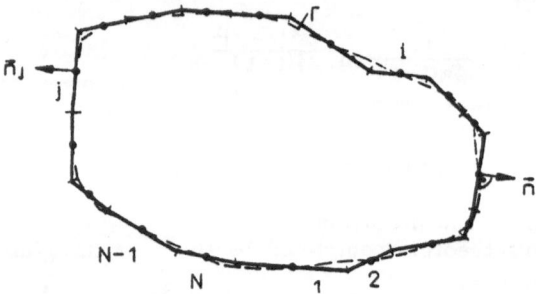

Figure 3. Discretization of the boundary of a plane problem with linear elements.

This geometrically discretization leads to a discretization of the boundary integral equation and if for example a plane magnetostatic field problem is considered, eq.(31) can be written as

$$c_i A_i + \sum_{j=1}^{N} \int_{\Gamma_j} A \frac{\partial G_{ij}}{\partial n_j} d\Gamma = \sum_{j=1}^{N} \int_{\Gamma_j} \frac{\partial A}{\partial n_j} G_{ij} d\Gamma + A_i^i \tag{110}$$

for the i-th field point (node) on the boundary. If only one node per element is assumed as shown in Fig.3, the field quantities A and $\partial A/\partial n$ are constant within the elements and eq.(110) becomes

$$c_i A_i + \sum_{j=1}^{N} A_j \int_{\Gamma_j} \frac{\partial G_{ij}}{\partial n_j} \, d\Gamma = \sum_{j=1}^{N} \frac{\partial A}{\partial n} j \int_{\Gamma_j} G_{ij} \, d\Gamma + A_i^i \quad . \tag{111}$$

Considering all nodes (i = 1,...,N) this leads to a system of linear algebraic equations which can be written as matrix equation

$$[H]\{A\} = [G]\left\{\frac{\partial A}{\partial n}\right\} + \{A^i\} \, , \tag{112}$$

where [H] and [G] are N x N - matrices, the so-called element matrices and $\{A\}$, $\{\partial A/\partial n\}$ are the two N - dimensional column vectors of the field quantities on the boundary. The vector $\{A^i\}$ represents the N known values of the impressed magnetic vector potential in the N nodes.

The coefficients of the element matrices are given as

$$[H]: \quad h_{ij} = \int_{\Gamma_j} \frac{\partial G(\mathbf{r}_i, \mathbf{r}_j)}{\partial n_j} \, d\Gamma + c_i \delta_{ij} \tag{113}$$

$$[G]: \quad g_{ij} = \int_{\Gamma_j} G(\mathbf{r}_i, \mathbf{r}_j) \, d\Gamma \tag{114}$$

where δ_{ij} is Kronecker's symbol

$$\delta_{ij} = \left\{ \begin{array}{l} 0 \text{ for } i \neq j \\ 1 \text{ for } i = j \end{array} \right. \tag{115}$$

The integrals in the equations (113) and (114) can be calculated either analytically or numerically. The acuracy of these integrations determines the accuracy of the solution of the equations system (112).

For potential problems the sum of the coefficients in each row of the [H] - matrix is zero, therefore, the diagonal elements can be calculated from the off-diagonal elements:

$$h_{ii} = -\sum_{\substack{i=1 \\ i \neq j}}^{N} h_{ij} \quad . \tag{116}$$

A higher accuracy of the results can be achieved by using shape functions of higher order for the approximation of the field quantities within the boundary elements. In this case multi-noded elements are necessary. For example Fig.4 shows one-dimensional shape functions of second order, which are used for the discretization of 2d problems with 3-noded elements. The parabolic shape functions

$$N_1 = \frac{1}{2}\xi(\xi - 1) \, , \tag{117}$$

$$N_2 = 1 - \xi^2 \, , \tag{118}$$

$$N_3 = \frac{1}{2}\xi(\xi + 1)$$

are normalized Lagrange polynomials [7].

44

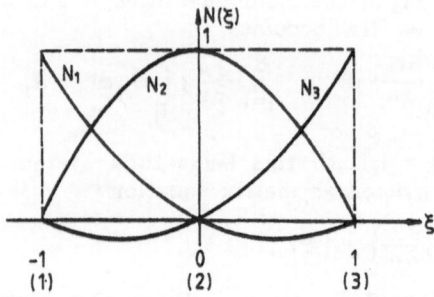

Figure 4. One-dimensional shape functions of second order.

The field quantities at any point of the boundary e.g. the magnetic vector potential and its normal derivative are calculated from the nodal values by linear combination of the equations (117), (118) and (119)

$$A(\xi) = \sum_{k=1}^{3} N_k(\xi) A_k \quad , \tag{120}$$

$$\frac{\partial A(\xi)}{\partial n} = \sum_{k=1}^{3} N_k(\xi) \frac{\partial A_k}{\partial n} \quad . \tag{121}$$

If so-called isoparametric elements are used the geometry is approximated the same way as the field quantities:

$$x(\xi) = \sum_{k=1}^{3} N_k(\xi) x_k \quad , \tag{122}$$

$$y(\xi) = \sum_{k=1}^{3} N_k(\xi) y_k \quad . \tag{123}$$

This approximation involves a transformation of a 3-noded parabolic element defined in the global coordinate system (x,y) into a linear element in the ξ-space, which is the interval [-1,1] as can be seen in Fig.5.

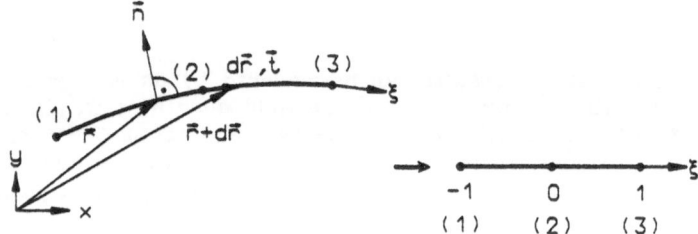

Figure 5. Transformation of a one-dimensional parabolic element.

The normal unit vector is defined as

$$n = \frac{1}{\sqrt{(\frac{dx}{d\xi})^2 + (\frac{dy}{d\xi})^2}} \left[-\frac{dy}{d\xi} e_x + \frac{dx}{d\xi} e_y \right] \tag{124}$$

where the derivatives are obtained from

$$\frac{dx}{d\xi} = \sum_{k=1}^{3} \frac{dN_k}{d\xi} x_k \, , \tag{125}$$

$$\frac{dy}{d\xi} = \sum_{k=1}^{3} \frac{dN_k}{d\xi} y_k \, . \tag{126}$$

The coefficients of the element matrices

$$h_{ij}^{(k)} = \int_{\xi=-1}^{1} N_k(\xi) \frac{\partial G(r_i, r_j(\xi))}{\partial n_j(\xi)} \, d\Gamma(\xi) \; + \; c_i \delta_{ij} \, , \tag{127}$$

$$g_{ij}^{(k)} = \int_{\xi=-1}^{1} N_k(\xi) \, G(r_i, r_j(\xi)) \, d\Gamma(\xi) \, , \tag{128}$$

with

$$d\Gamma(\xi) = \sqrt{\left(\frac{dx}{d\xi}\right)^2 + \left(\frac{dy}{d\xi}\right)^2} \; d\xi \tag{129}$$

may be calculated for example numerically by a Gaussian quadrature. The index k indicates the local node number of the considered source point j, which may be 3 in one element and 1 in the neighbouring element. If one node belongs to several elements the coefficients in the equations (127) and (128) have to be assembled for the compilation of the element matrices [H] and [G], e.g.:

$$h_{ij} = h_{ij}^{(3)}\Big|_{\text{element } (l-1)} + h_{ij}^{(1)}\Big|_{\text{element } l} \, , \tag{130}$$

$$g_{ij} = g_{ij}^{(3)}\Big|_{\text{element } (l-1)} + g_{ij}^{(1)}\Big|_{\text{element } l} \, . \tag{131}$$

For the discretization of the boundary of a 3d problem quadrilateral elements of second order may be used. In Fig.6 the transformation from the global to the local coordinate system of such an element is shown.

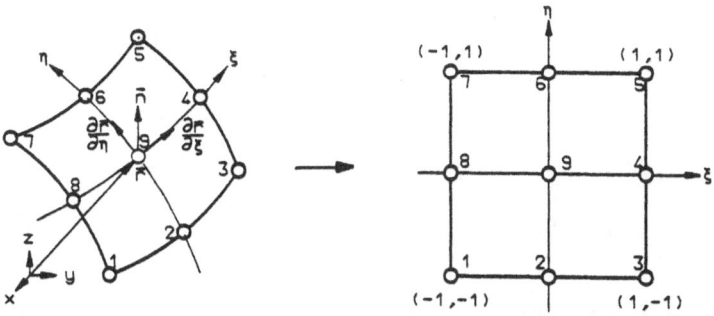

Figure 6. Transformation of a two-dimensional parabolic element.

For the application of complete 2d shape functions of second order 9-noded elements are necessary. In order to minimize the number of nodes usually the central node is neglected which means that incomplete shape functions of second order have to be used [7].

In our software packages a rough structure (macro structure) is defined with 9-noded elements which can be automatically subdivided into 8-noded elements for a finer mesh. the field quantities and the geometry are approximated with

$$\Phi(\xi,\eta) = \sum_{k=1}^{8\,or\,9} N_k(\xi,\eta)\, \Phi_k \qquad , \tag{132}$$

$$\frac{\partial\,\Phi(\xi,\eta)}{\partial\,n} = \sum_{k=1}^{8\,or\,9} N_k(\xi,\eta)\frac{\partial\,\Phi}{\partial\,n}k \qquad , \tag{133}$$

$$x(\xi,\eta) = \sum_{k=1}^{8\,or\,9} N_k(\xi,\eta)\, x_k \qquad . \tag{134}$$

The normal unit vector is defined as

$$n = \frac{\dfrac{\partial\,r}{\partial\,\xi} \times \dfrac{\partial\,r}{\partial\,\eta}}{\left|\dfrac{\partial\,r}{\partial\,\xi} \times \dfrac{\partial\,r}{\partial\,\eta}\right|} \tag{135}$$

and the differential surface element can be calculated by

$$d\Gamma(\xi,\eta) = \left|\frac{\partial\,r}{\partial\,\xi} \times \frac{\partial\,r}{\partial\,\eta}\right| d\xi\,d\eta \qquad . \tag{136}$$

With the equations (132) until (136) the non-assembled coefficients $h_{ij}^{(k)}$ and $g_{ij}^{(k)}$ can be calculated (e.g. numerically) by Gaussian quadrature in the ξ,η-coordinate sytem. Care must be taken for the calculation of the diagonal elements because the integral kernels become singular. The coefficients h_{ii} can be calculated from the off-diagonal elements. For the calculation of the coefficients g_{ii} a sub-division technique shown in Fig.7 is used in our software package.

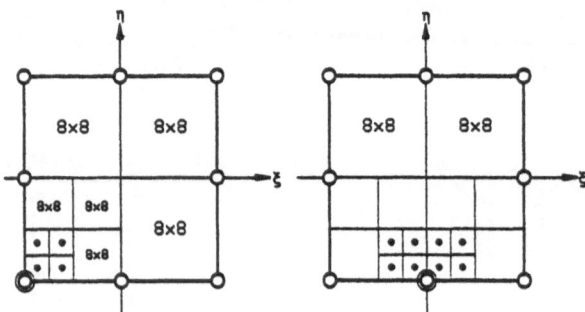

Figure 7. Subdivision technique for the calculation of integrals with singular kernels.

From Fig.7 it can be seen that the subdivisions are concentrated to the respective singular point. Within the subdivisions a constant number of Gauss points is used.

After rearranging eq.(112) where the known boundary values of A or $\partial A/\partial n$ are collected on the right-hand side the remaining system of linear algebraic equations

$$[A]\{X\} = \{b\} \tag{137}$$

must be solved. As well direct as iterative methods are applicable for the solution of an equations system. For integral methods the coefficient matrix [A] is in general an unsymmetric full populated matrix, which is relatively small. In this case the application of a direct method, e.g. the Gauss elimination method seems to be preferable. For larger coefficient matrices comprising many zero elements iterative methods are preferable. For example the application of the Finite Element Method leads to a very large sparse matrix [A] where the equations system can be solved very effectively with the conjugate gradient method.

4. APPLICATION

In this chapter a few examples shall illustrate the application of the BEM using various formulations for the solution of magnetostatic and eddy current problems [8], [9]. The numerical results are compared with solutions obtained by FEM.

4.1. C - SHAPED MAGNET

In this example a 3d magnetostatic field problem is studied. The problem consists of a C-shaped iron core ($\mu_r = 100$) with a large air gap, which is excited by two cylindrical coils symmetrically arranged. This problem can be considered as an open problem with two domains, i.e. the iron core and the surrounding air region including the coil systems. The calculations are based on a total and reduced scalar potential formulation (eq.(48)) considering the interface conditions (54) and (55).

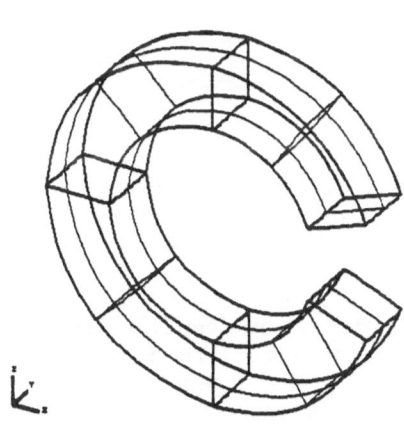

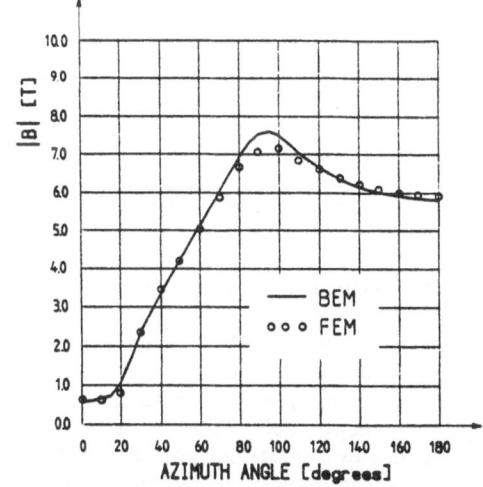

Figure 10. Boundary element mesh of a C-shaped magnet.

Figure 11. Magnitude of the magnetic flux density as function of the azimuth angle.

In Fig.10 the boundary element mesh of the magnet's surface is shown, representing the interface between iron and air domains. For the numerical computations a discretization with 52 eight-noded elements resulting in 158 nodes was used. As solution of the equations system the total scalar potential and its normal derivative are obtained in each node on the boundary. The magnetic flux density within the iron core is calculated by using the equations (4) and (52) and the integral representation for the total scalar potential. In Fig.11 the numerical results for the magnetic flux density along a central line as function of the azimuth angle is presented.

4.2. IRON CYLINDER

This example illustrates the behaviour of an isotropic ($\mu_r = 100$) and an orthotropic ($\mu_{rx} = 10$, $\mu_{ry} = 10$, $\mu_{rz} = 100$) cylinder, respectively, which is immersed in the magnetic field of an cylindrical coil of the same height as the iron core. This problem again is solved by an total and reduced scalar potential formulation, whereby for the orthotropic case the Green's function of eq.(61) is in effect.

In Fig.12 the boundary element mesh of the iron cylinder consisting of 26 elements with 80 nodes in total is shown. In Fig.13 the magnetic flux density in the symmetry plane versus the radial distance inside and outside the iron core is plotted.

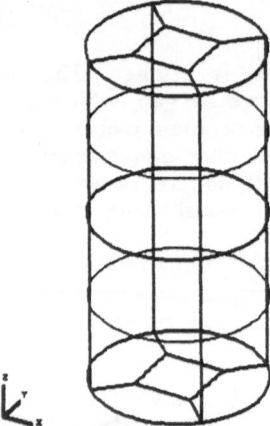

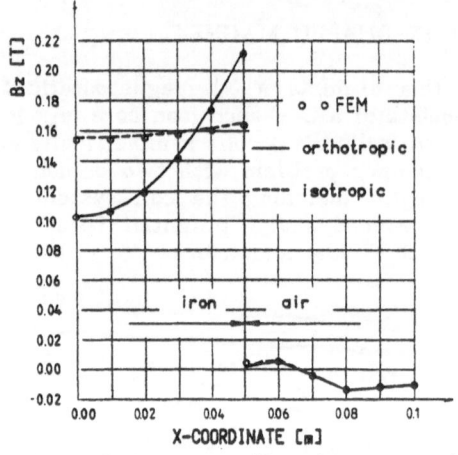

Figure 12. Boundary element mesh of the cylindrical core.

Figure 13. Magnetic flux density in the symmetry plane along the x-axis.

4.3. ASYMMETRICAL CONDUCTOR WITH A HOLE

In this problem, which is a current TEAM Workshop problem [10], the eddy current distribution in a thick aluminum plate with an excentrical hole is considered. The plate is placed unsymmetrically in the magnetic field of a racetrack shaped coil. This problem represents a multi-connected field problem. The governing equations are the coupled boundary integral equations (79) and (80) for the magnetic vector potential and the electric scalar potential, respectively. In order to apply the BEM the surface of the metal plate is divided into 8-noded quadrilateral elements. The impressed magnetic vector potential of the

exciting coil is calculated by analytical and numerical integration.
In Fig.14 a grey shaded image of the magnitude of the eddy current density at
the surface of the conducting plate is shown. For this example a harmonic time
variation of 50 Hz was assumed. The calculations were performed with a
discretization of the plate into 96 boundary elements and a total of 288 nodes.

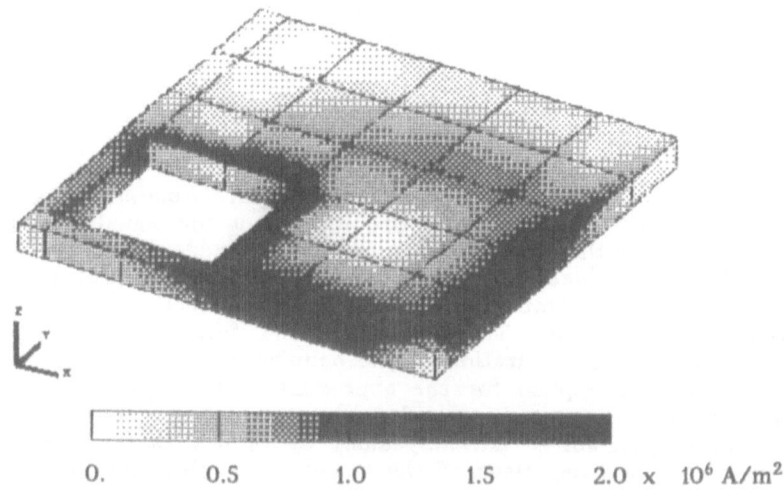

0.   0.5   1.0   1.5   2.0 x 10^6 A/m^2

Figure 14. Magnitude of the eddy current density for f = 50 Hz.

4.4. FELIX LONG CYLINDER EXPERIMENT

This problem was also defined as a test problem for the TEAM workshops.

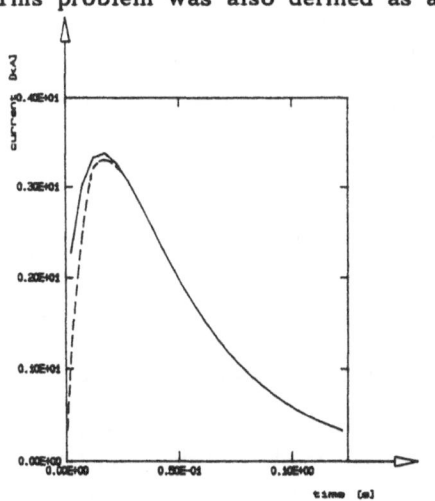

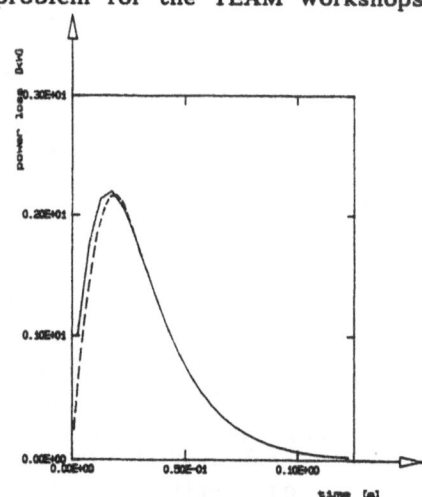

Figure 15. Total eddy current
of the FELIX cylinder.
(—— BEM, - - - - FEM)

Figure 16. Power losses in the
wall of the FELIX cylinder.
(—— BEM, - - - - FEM)

A thin walled long hollow aluminum cylinder (inner radius: 0.1317 m, outer radius: 0.1365 m) is situated in a uniform magnetic field. The magnetic field is directed perpendicular to the axis of the cylinder and decays exponentially in time ($B_0 = 0.1$ T, $\tau = 0.0397$s).

The 2d BEM calculations are based on the time stepping method (eq.(102)) with constant time steps of $\Delta t = 0.005$ s. The computations were performed with 40 one-noded linear elements for both the inner and the outer boundary. In the Figures 15 and 16 the time dependence of the total eddy current and the power losses are shown.

5. CONCLUSION

A general overview was given of various integral formulations for the solution of magnetostatic and eddy current problems using the boundary element method (BEM). The resulting relationships can be extended to many other electromagnetic field problems. It was pointed out that for the calculation of three-dimensional magnetostatic field problems the application of a magnetic scalar potential is preferable. The presented BEM formulation uses quadrilateral elements for the discretization of the boundaries, whereby shape functions of higher order are applied for the approximations of the solution and of the geometry. The presented examples show that many electromagnetic field problems can be solved advantageously by BEM with high accuracy. This is demonstrated by comparison of the numerical results with FEM solutions.

References

[1] Harrington, R.F. (1968) Field computation by moment methods, The Macmillan Company, New York

[2] Brebbia, C.A. and Walker S. (1980) Boundary element techniques in engineering, Newnes-Butterworths, London.

[3] Simkin, J. and Trowbridge C.W. (1979) 'On the use of the total scalar potential in the numerical solution of field problems in electromagnetics', Int. J. Num. Meth. Eng., vol. 14, 423-440.

[4] Rucker, W.M. (1985) 'Boundary element calculation of magnetostatic field problems using the reduced and total scalar potential', Proceedings 1st IGTE symposium Graz, 38-46.

[5] Kalaichelvan S. and Lavers J.D. (1989) 'Boundary element methods for eddy current problems', Topics in Boundary Element Resarch (Ed. C.A. Brebbia), vol. 6, Springer-Verlag, 78-117.

[6] Morse, P.M. and Feshbach H. (1953) Methods of Theoretical Physics, McGraw-Hill, New York.

[7] Cheung, Y.K. and Yeo M.F. (1979) A practical introduction to finite element analysis, Pitman, London.

[8] Rucker, W.M. and Richter K.R. (1988) 'Three-dimensional magnetostatic field calculation using boundary element method', IEEE Transactions on Magnetics, vol. 24, 23-26.

[9] Rucker, W.M. and Richter K.R. (1990) 'A BEM code for 3-d eddy current calculations', IEEE Transactions on Magnetics, vol. 26, 462-465.

[10] Turner, L. (Editor) (1988) TEAM Workshops: Testproblems.

DIFFERENTIAL METHODS, FINITE ELEMENTS AND APPLICATIONS

D. RODGER
School of Electrical Engineering
University of Bath
Claverton Down
Bath BA2 7AY, UK

ABSTRACT.   Differential finite element methods for calculating power frequency 3-dimensional electromagnetic field distributions are described.   The problem region is first partitioned into regions which allow no eddy currents, and regions in which eddy currents flow.   Magnetic scalars are used to describe the field in non-eddy regions.   The magnetic vector potential A, with or without an auxiliary electric scalar potential V is used in eddy regions.   Variations on this basic formulation which allow moving conductors, time transient fields and multiply connected regions to be supported are described.   An economic method for dealing with eddy currents which flow in thin sheets is also outlined.

1.   Introduction

At present many different approaches to formulating and solving general 3D electromagnetic field problems are being investigated worldwide.   An appreciation of the scope and diversity of the on-going work may be readily gained by consulting past COMPUMAG [1] and INTERMAG [2] conference proceedings.   Although the most frequently used techniques differ in detail, usually they can be broadly categorised according to how fields are modelled inside eddy current regions, as follows:

1.   Methods which use E the electric field or a potential such as A the magnetic vector potential.

2.   Methods which use H the magnetic field intensity or a potential such as T the electric vector potential.

Auxiliary scalar potentials may also be used in conjunction with the vector potentials.   The most economic implementations use magnetic scalar potentials to model fields in non conducting regions.   This can give rise to special problems when these regions become multiply connected, as discussed in section 7.

Both methods 1 and 2 are commonly implemented using finite elements of either the 'nodal variable' or the 'edge variable type'.   This gives rise to four different types of implementation.   More permutations than this are possible if we consider that it is possible to solve either the differential or the integral forms of the field equations (or, indeed, a mixture of both).

51

Y. R. Crutzen et al. (eds.), Industrial Application of Electromagnetic Computer Codes, 51–79.
© 1990 *ECSC, EEC, EAEC, Brussels and Luxembourg.*

In addition to the above techniques which are intended to be used for general problems, there are a number of 'special' techniques which are very important, for example, for the solution of eddy currents in thin sheets or for dealing with fine cracks.

In this contribution we will only consider the differential form of the first of the above general methods; details of the other methods are dealt with elsewhere.

2. Derivation of the Field Equations

The volume in which a field solution is required is first partitioned into regions which contain no eddy currents and regions in which eddy currents flow in such a way that the dividing surfaces are the surfaces of the conductors. The whole volume is then discretised according to the standard finite element methodology.

Magnetic scalar potentials are then used to describe the field in non–eddy regions and a three–component vector A, the magnetic vector potential, either with or without an electric scalar potential V, is used in eddy regions. This is described in the following few sections:

2.1 MAXWELL'S EQUATIONS AND BASIC ASSUMPTIONS

The relationship between the various field quantities at power frequency (the quasistationary limit) are given by a subset of Maxwell's equations:

$$\text{curl } H = J \tag{1}$$

$$\text{curl } E = -\frac{\partial B}{\partial t} \tag{2}$$

$$\text{div } B = 0 \tag{3}$$

$$\text{div } J = 0 \tag{4}$$

The material properties are defined by:

$$B = \mu H \quad (\text{or } H = \nu B) \tag{5}$$

$$J = \sigma E \tag{6}$$

Fig. 1 depicts a representative eddy–current problem which has already been partitioned into three types of region after the manner discussed at the beginning of Section 2. The different variables which describe the field in each volume are shown. Region 1 contains all the source currents and no iron, region 2 contains no currents at all and region 3 represents the conductor.

2.2 FIELD EQUATIONS FOR THE NONCONDUCTING REGIONS

In region 1 of Fig. 1 the field H_s due to the source currents alone may easily be calculated using the Biot–Savart law, since

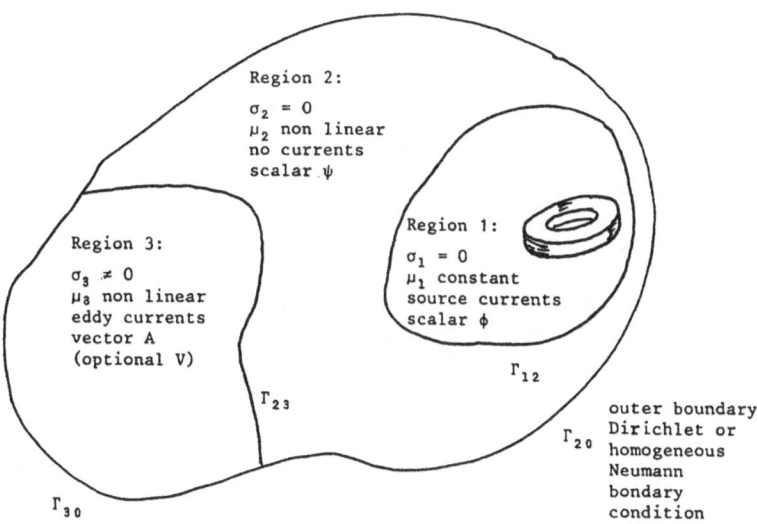

Fig. 1   Partitioning for a typical eddy-current problem

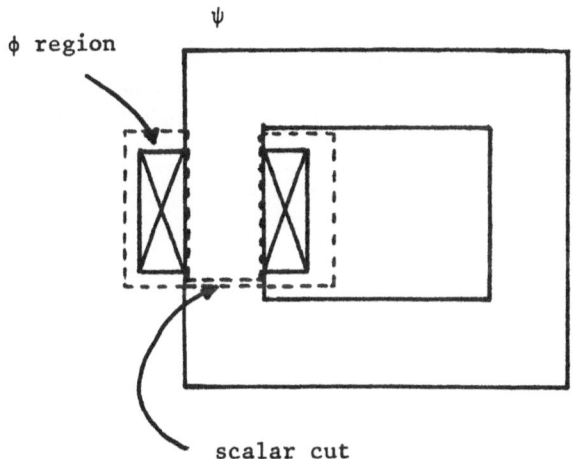

Fig. 2   Showing a cut through the total magnetic scalar ψ for a transformer problem

$$\text{curl } H_s - J_s \tag{7}$$

If a field H_d is defined as

$$H_d - H_1 - H_s \tag{8}$$

where H_1 is the total field in region 1, as curl H_1 is equal to curl H_s then curl H_d is zero. This means that H_d can be described as the gradient of a scalar φ, as curl grad φ is always zero for any scalar φ.

The Helmholtz theorem states that all vector fields are completely described if the divergence and curl throughout the volume, together with the normal component over the enclosing surface, are known. This means that H_1 can be obtained by solving only eqn. 3; that is, from eqn. 3 and upon substituting φ into eqn. 8:

$$\text{div } \mu_1 \ H_1 - 0 - -\text{div } \mu_1 \text{ grad } \varphi + \text{div } \mu_1 \ H_s \tag{9}$$

This was the form of one of the first finite–element solution methods which used magnetic scalar potentials for solving 3–dimensional magnetostatic field problems [3]. However, it was pointed out [4] that obtaining the total field in this way, when eqn. 9 is applied to a volume containing iron, will lead to large errors. Inside iron, H_d and H_s are nearly oppositely directed, as $\mu \rightarrow \infty$, $H_1 \rightarrow 0$. This difficulty was circumvented [4] by allowing no iron in region 1 and describing the field in regions which can contain iron (region 2 in Fig. 1) as the gradient of a second scalar:

$$H_2 - -\text{grad } \psi \tag{10}$$

This is possible, because no currents exist in region 2. An equation similar to eqn. 9 defines H_2. It was then shown how magnetostatic field problems can be solved in terms of two scalar potentials φ and ψ. A very convenient scheme was introduced whereby the interface conditions (eqns. 21 and 22) applied to the surface Γ_{12}':

$$-\mu_1 \frac{\partial \varphi}{\partial n} + \mu_1 \ H_{sn} - \mu_2 \frac{\partial \psi}{\partial n} \tag{11}$$

and

$$- \frac{\partial \varphi}{\partial r} + H_{sr} - - \frac{\partial \psi}{\partial r} \tag{12}$$

were used to eliminate either φ or ψ on the surface, where r represents a tangent to Γ_{12}. H_s need therefore only be calculated on Γ_{12}.

It is interesting to note that the source coil regions can also become multiply connected. This happens in situations where iron passes through a coil, as in transformers. The problem is quite easily solved, by allowing the reduced scalar potential region to exist on a 'cut' which spans the total scalar region, as shown in Fig. 2.

2.3 FIELD EQUATIONS FOR THE CONDUCTING REGIONS

It is worthwhile to consider first of all the case where the conductivity σ of a conducting region is constant. This is a fairly common case and gives rise to an important saving in computer time, as the electric scalar potential V is not required.

2.3.1. *Derivation of B from a Magnetic Vector Potential, Constant Conductivity*: With reference to Fig. 1, region 3 represents a volume in which eddy currents can flow.

A classical method for finding B involves first defining a magnetic vector potential, such that:

$$\text{curl } A = B \tag{13}$$

Any field B found from A, as defined in eqn. 13, will have zero divergence because of the vector identity div curl A equals zero for any A, so that eqn. 3 is satisfied and, by the Helmholtz theorem, we are nearly half way towards defining B.

From eqns. 2 and 13 we have

$$\text{curl } E = - \frac{\partial}{\partial t} \text{curl } A \tag{14}$$

so that, integrating eqn. 14,

$$E = - \frac{\partial A}{\partial t} - \text{grad } V \tag{15}$$

In the general case, an electric scalar potential V can exist in eqn. 15, as taking the curl of eqn. 15 yields eqn. 14 for any scalar V, as curl grad V is always zero. It will be shown that, for this particular case of constant σ, V can be allowed to be zero everywhere inside region 3. At this stage an equation involving the curl of A can be derived from eqns. 1, 13 and 15.

$$\text{curl } \nu \text{ curl } A = -\sigma \left[\frac{\partial A}{\partial t} + \text{grad } V \right] \tag{16}$$

This equation leaves the divergence of A unspecified, so that by the Helmholtz theorem it is not enough to define a unique A. It is worthwhile, therefore, to examine div A.

2.3.2 *The Divergence of A and E:* Eqn. 15 arises from the fact that div J is always zero (at power frequencies) while J is equal to σE and σ can be discontinuous. Consequently, E has divergence sources at any discontinuity in σ, that is,

$$\text{div } J = \text{div } \sigma E = 0 = \sigma \text{ div } E + E \cdot \text{grad } \sigma \tag{17}$$

so that

$$\text{div } E = -\frac{1}{\sigma} E . \text{grad } \sigma \tag{18}$$

The main reason for computing E via A and V has always been that simpler choices of divergence can be picked for A than for E, as that of A can to some extent be arbitrary. As:

$$\text{div } \sigma E = 0 = -\text{div}\left[\sigma\frac{\partial A}{\partial t}\right] -\text{div }(\sigma \text{ grad } V) \tag{19}$$

any convenient choice of div A will suffice, so long as V is simultaneously calculated such that the right-hand side of eqn. 19 always balances to give zero.

This can be appreciated most easily by examining the conditions at an interface between a conductor and air. Just inside the conductor J_n is zero (eqn. 4) and therefore E_n is zero also (eqn. 6). Just outside the conductor, however, E_n is nonzero, so that E_n is discontinuous. This is a problem, as discontinuous field quantities are usually inconvenient in numerical representations of fields. (A description of how this can be carried out for 'reasonably smooth' interfaces between regions of discontinuous conductivity can be found in ref [5]). A procedure which allows a discontinuous E to be modelled using a continuous field variable (A) can be implemented by calculating A outside the conductors, together with A and, simultaneously, V inside in such a way that

$$-\frac{\partial V}{\partial n} - \frac{\partial A_n}{\partial t} = E_n = 0 \tag{20}$$

on the inner conductor surfaces and E_n is equal to $-\partial A_n/\partial t$ outside.

As A is continuous, obtaining a zero E_n by directly setting A_n to zero just inside the conductors while ignoring V would result in the wrong value, zero, being obtained for E_n outside the conductor in this case. The above choice of V (eqn. 20) associates it directly with the surface charges and most numerical 3-dimensional solutions involving A have proceeded along these lines, although many other choices exist. A very good discussion of this is available in ref [6].

In the present formulation, the region exterior to conductors is modelled by means of a magnetic scalar potential, and this results in a surface discontinuity in E being readily modelled. That is, E_n can be directly set to zero on the inside of conductor surfaces, making div E zero (σ is constant) throughout the region of interest. To show how this simple option can exist, we now examine the interface conditions between regions 2 and 3 of Fig. 1.

The continuity conditions resulting from Maxwell's equations are:

$$n . B_2 = n . B_3 \tag{21}$$

$$n \times H_2 = n \times H_3 \tag{22}$$

$$n \times E_2 = n \times E_3 \tag{23}$$

$$n.J_3 = 0 \tag{24}$$

Substituting the relevant variables in eqn. 21 yields

$$-\mu_2 \frac{\partial \psi}{\partial n} = n . \; curl \; A \tag{25}$$

Similarly, for eqn. 22

$$-n \times grad \; \psi = \nu_3 \; curl \; A \times n \tag{26}$$

Eqn. 23, describing the continuity of tangential E, can be ignored in this case, because E has no effect on the magnetic field outside conductors. Eqn. 24 can be dealt with directly by setting E_n to zero. All the continuity conditions are therefore readily satisfied, while retaining the simple (zero) divergence for A and E. Eqns. 25 and 26 will be further discussed in Section 3.

Now that we have E equal to $-\partial A/\partial t$ and the divergence of A and E both equal to zero, the usual reason for using A instead of E as a field variable (simpler divergence) has disappeared. There are two remaining reasons for using A instead of E. D.C. conditions can be modelled using A, also a symmetric system of equations is easier to obtain when using the former, owing to the $\partial/\partial t$ term which is present in eqn. 2 but not in eqn. 13. In the original implementation of this method [7] E was used.

In summary then, when dealing with constant σ, it is necessary to solve:

$$curl \; \nu \; curl \; A = -\sigma \frac{\partial A}{\partial t} \tag{27}$$

together with div μ grad $\psi = 0$.

2.3.3 *Discontinuous Conductivity:* When σ is discontinuous and the interfaces between the different conducting regions in contact are complicated, we must resort to the electric scalar potential. Restating eqn (16), we have:

$$curl \; \nu \; curl \; A = -\sigma \left[\frac{\partial A}{\partial t} + grad \; V \right] \tag{28}$$

This comprises three equations and four unknowns, so obviously cannot be solved. If we add:

$$div \; \sigma \left[\frac{\partial A}{\partial t} + grad \; V \right] = 0 \tag{29}$$

as our fourth equation, we will obtain a symmetric system of equations, which is important, but we have not added any new information since eqn (29) can be obtained by taking the divergence of equation (28). (div of the left hand side must always be zero, as div curl (vector) is always zero.)

The fourth piece of information which we need is that the divergence of **A** must be specified (from Helmholtz). Neither eqns 28 nor 29 specify the divergence of **A**. Eqn. 29 states that

$$\mathrm{div} \left[\sigma \frac{\partial A}{\partial t} \right]$$

can be anything so long as div σ grad V is equal and opposite.

Eqn (27) also does not specify the divergence of **A** in cases where the right hand term is relatively small. (Again, taking the divergence of eqn (27) shows that div (left side) is always zero, so that

$$\mathrm{div} \left[\sigma \frac{\partial A}{\partial t} \right]$$

is zero also. This is sufficient to 'pin down' the divergence of **A** if σ $\partial/\partial t$ is sufficiently large. However, the loss of definition of the divergence of **A** in a steady state a.c. problem can be readily observed as the frequency is reduced.

In the general case, then, it is necessary to specify more information about the divergence of **A**. This has been done by a variety of methods, which will be described in section 3.5.

3. Numerical Solution of the Field Equations

3.1 GENERAL METHOD

To find an approximate solution to the set of coupled field equations which describe regions 1–3 of Fig. 1, a finite–element technique, the Galerkin weighted residual method, can be used [8]. The implementation described here will use standard 'nodal variable' finite elements, as stated already 'edge variable' finite elements are also possible. The latter have the advantage that the divergence of the field quantitites are easier to specify, but sometimes accuracy is reduced [9]. Regions 1–3 of Fig. 1 will now be discussed separately.

3.2 NUMERICAL MODEL OF THE NONCONDUCTING REGIONS

Only region 2 and its interface with region 3 will be discussed, as region 1 is very similar. Further, the interface conditions on $\Gamma_{1\,2}$ will be omitted, as these have been well documented [4].

Applying a weighted residual procedure to the Laplacian equation in ψ results in a set of linear equations of the form

$$\int \omega_i (\mathrm{div} \; \mu_2 \; \mathrm{grad} \; \psi) \; d\Omega - 0 \tag{30}$$

in which ω_i is a suitable set of m weighting functions. It is convenient to transform this equation by means of Green's theorem, as this avoids having second derivatives in the functional:

$$\int \mu_2 \ grad \ \omega_i \ . \ grad \ \psi \ d\Omega \ - \oint_{r_2} \mu_2 \omega_i \ \frac{\partial \psi}{\partial n} \ d\Gamma = 0 \qquad (31)$$

The scalar field volume is now divided up into a set of simple subdomains (finite elements). It is assumed that the field ψ can be adequately represented by a finite set of m values ψ_i situated at the nodes of the finite–element mesh.

Within each finite element the field ψ is then described by a linear combination of the nodal values (ψ_i) and simple geometric functions of the co–ordinates, shape functions.

$$\psi = \sum_{i=1}^{k} N_i \psi_i \qquad (32)$$

where k is the number of element nodes.

In the Galerkin procedure the weighting functions ω_i are the same as the shape functions N_i. In eqn. 31, the field ψ can now be expanded according to eqn. 32 for each finite element, so that the m linear equations (eqn. 31) are formed by summing local contributions on an element by element basis. Thus, the contribution for the ith node of element e is

$$\int_{\Omega_e} \mu_e \sum_{j=1}^{k} \left[\frac{\partial N_i}{\partial x} \frac{\partial N_j}{\partial x} + \frac{\partial N_i}{\partial y} \frac{\partial N_j}{\partial y} + \frac{\partial N_i}{\partial z} \frac{\partial N_j}{\partial z} \right] \psi_j \ d\Omega$$

$$- \oint_{\Gamma_e} \mu_e N_i \ \frac{\partial \psi}{\partial n} \ d\Gamma \qquad (33)$$

The surface term in eqn. 33 can be written

$$\oint_{\Gamma_e} \mu_e N_i \ \frac{\partial \psi}{\partial n} \ d\Gamma = \sum_{p=1}^{s} \int_{\Gamma_{ep}} \mu_e N_i \left[\frac{\partial \psi}{\partial n} \right]_p d\Gamma \qquad (34)$$

where s is the number of facets belonging to element e. If Γ_{ep} is an interior surface of region 2, ignoring the surface integral in the summation will automatically yield the continuity condition of eqn. 21, as the same integral with an opposite sign for the normal n will arise from the other element which shares Γ_{ep}. In the same way, if Γ_{ep} is on Γ_{20}, then a homogeneous Neumann condition arises naturally if the surface term is ignored.

When Γ_{ep} is on Γ_{23}, the surface term must be calculated and is used to link the normal component of B between region 2 and region 3, that is,

$$- \int_{\Gamma_{ep}} \mu_e N_i \left[\frac{\partial \psi}{\partial n} \right]_p d\Gamma = \int_{\Gamma_{ep}} N_i n \ . \ curl \ A \ d\Gamma \qquad (35)$$

where curl **A** must be expanded in terms of the shape functions and nodal values of **A** from the adjacent element in region 3.

3.3 NUMERICAL MODEL OF THE CONDUCTING REGIONS WITHOUT THE ELECTRIC SCALAR V

3.3.1 *Basic Equation in A*: Applying the Galerkin method to eqn. 16 (with V zero) results in:

$$\int_{\Omega_e} N_i \cdot \left[\text{curl } \nu \text{ curl } A + \sigma \frac{\partial A}{\partial t} \right] d\Omega = 0 \tag{36}$$

where N_i is now a vector shape function:

$$N_i = N_{xi} i + N_{yi} j + N_{zi} k \tag{37}$$

where $N_{xi} = N_{yi} = N_{zi}$ in this case. The field **A** is described as:

$$A = \sum_{i=1}^{k} N_i A_i \tag{38}$$

in a finite element with k nodes.

Eqn. 36 is now treated in two different ways, depending on whether ν is constant throughout region 3.

3.2.2 *Constant ν, the Vector Poisson Equation*: If ν is constant, the first term in eqn. 36 can be expanded as:

$$\text{curl } \nu \text{ curl } A = -\nu \text{ div grad } A + \nu \text{ grad div } A \tag{39}$$

Setting div **A** to zero and substituting eqn. 39 into eqn. 36 leads to:

$$\int_{\Omega_3} N_{xi} \left[-\nu \text{ div grad } A_x + \sigma \frac{\partial A_x}{\partial t} \right] d\Omega \tag{40}$$

with similar equations for y and z.

Again, applying Green's theorem as for eqn. 30 yields:

$$\int_{\Omega_3} \left[\nu \text{ grad } N_{xi} \cdot \text{grad } A_x + N_{xi} \sigma \frac{\partial A_x}{\partial t} \right] d\Omega$$

$$- \oint_{\Gamma_e} N_{xi} \, \nu \, \frac{\partial A_x}{\partial n} \, d\Gamma = 0 \tag{41}$$

for node i at element e, for the N_x equation. The equations for N_{yi} and N_{zi} are of exactly the same form.

3.3.3 *Non-Linear v, the Curl Curl Equation*: Applying Green's theorem to eqn. 33 yields:

$$\int_{\Omega_3} \nu \ \text{curl} \ N_i \ . \ \text{curl} \ A + \sigma N_i \ . \ \frac{\partial A}{\partial t} \ d\Omega$$

$$- \oint_{\Gamma_3} N_i \ . \ (\nu \ \text{curl} \ A \ x \ n) \ d\Gamma = 0 \tag{42}$$

Again, the surface term for eqn. 42 may be expressed as a summation of surface integrals over all the facets of each element, as in eqn. 34. Also, a similar argument can be used as for the ψ equation, to show that for all interior facets of region 3 the surface terms ensure continuity of the tangential component of H (eqn. 22). This is also true in the case of the surface term in eqn. 41, but only when ν is constant throughout all the conductor; this can readily be appreciated by examining the interface conditions at a surface on which n = i. We know that H_y and H_z must be continuous. Continuity of H_y requires that

$$-\nu_1 \ \frac{\partial A_z}{\partial x_1} + \nu_1 \ \frac{\partial A_x}{\partial z_1} = -\nu_2 \ \frac{\partial A_z}{\partial x_2} + \nu_2 \ \frac{\partial A_x}{\partial z_2} \tag{43}$$

Since $\partial A_x/\partial z$ is continuous (the transverse derivatives of a continuous function are always continuous), eqn. 43 indicates that there is a discontinuity in the normal derivative of A_z when $\nu_2 \neq \nu_1$, that is

$$\frac{\partial A_z}{\partial x_1} - \frac{\nu_2}{\nu_1} \frac{\partial A_z}{\partial x_2} + \frac{(\nu_1 - \nu_2)}{\nu_1} \frac{\partial A_x}{\partial z} \tag{44}$$

This condition is natural to the curl curl functional in eqn. 42, but is only natural to the vector Poisson functional of eqn. 41, when $\nu_2 = \nu_1$. Simple continuity of the A components is enough to guarantee continuity of B_n throughout the conductor volume, and so this is correct for both functionals.

On Γ_{23} the surface term in eqn. 41 may be expanded as:

$$\int_{\Gamma_{ep}} N_{xi} \ \left[\ \nu \ \left[\frac{\partial A_y}{\partial y} + \frac{\partial A_z}{\partial z} \right] \ 1_x - \left[\frac{\partial \psi}{\partial z} + \nu \ \frac{\partial A_y}{\partial x} \right] \ 1_y \right.$$

$$\left. + \left[\frac{\partial \psi}{\partial y} - \nu \ \frac{\partial A_z}{\partial x} \right] \ 1_z \right] \ d\Gamma \tag{45}$$

in which the first two terms arise from div $\mathbf{A} = 0$ and where l_x, l_y, l_z are the direction cosines of the normal n to the facet. Similarly, in eqn. 42

$$\int_{\Gamma_{ep}} \mathbf{N} \cdot (\nu \text{ curl } \mathbf{A} \times n) \, d\Gamma - \int_{\Gamma_{ep}} \mathbf{N} \cdot (-\text{grad } \psi \times n) \, d\Gamma \qquad (46)$$

3.4 NUMERICAL MODEL OF THE CONDUCTING REGIONS WITH THE SCALAR V

The Galerkin technique may be applied to eqns. (28) and (29). As described before, in section 3.3, a 'vector Poisson' or 'curl curl' version of the A terms are possible. In this section, for brevity, only the more general curl curl version will be examined.

Thus we have, from eqns. (28) and (29), after transformation:

$$\int_{\Omega_3} \nu \text{ curl } N_i \cdot \text{curl } \mathbf{A} + \sigma N_i \cdot \left[\frac{\partial \mathbf{A}}{\partial t} + \text{grad } V \right] d\Omega$$

$$- \oint_{\Gamma_3} N_i \cdot (\nu \text{ curl } \mathbf{A} \times n) \, d\Gamma = 0$$

and

$$\int_{\Omega_3} \sigma \text{ grad } N_i \cdot \left[\frac{\partial \mathbf{A}}{\partial t} + \text{grad } V \right] d\Omega - \oint_{\Gamma_3} \sigma N_i \left[\frac{\partial \mathbf{A}}{\partial t} + \text{grad } V \right] \cdot n d\Gamma = 0$$

$$(47)$$

Equation (47) can be rendered symmetric if we use a variable $\partial v / \partial t$ in place of V.

3.5 IMPOSING THE DIVERGENCE OF A

As discussed, the divergence of A has to imposed in the general case. Any convenient gauge may be chosen, the most convenient is that div $\mathbf{A} = 0$. This, combined with the condition that $\mathbf{A}.n = 0$ on the inside surfaces of conductors, ensures a unique A. This is the Coulomb gauge. Other schemes, for instance the Lorentz gauge, appear in the literature [10,11].

The condition div $\mathbf{A} = 0$ can be imposed in several different ways, the most convenient of which is either to impose the condition directly using Lagrange multipliers [12] or to use a penalty technique [13]. The former technique involves introducing more unknowns into the final set of equations, so we now use the latter in the MEGA program.

The penalty technique is well known in the finite element literature. Several different strategies can be followed which lead to the same result:

1. The term $\alpha(\text{div } \mathbf{A})^2$ can be added to the functional associated with the curl ν curl A equation.

2. The term α div N div A can be added to the usual Galerkin form (eqn. 47).

3. The term grad $1/\mu_0$ div A has been added to eqn. (28) [14].

All three starting points lead to exactly the same terms in the final matrix.
The last method is equivalent if α, the large penalty number, is equal to $1/\mu_0$.

4. A Historical Example – The Bath Cube

For nostalgic reasons, it now seems appropriate to impose some early results on the gentle reader. One of the first 3D eddy current experiments consisted of blocks of aluminium exposed to time sinusoidally varying magnetic fields between the poles of a magnet [15].

The apparatus and three test cases are described and sketched on Fig. 3. The finite element results shown here are from ref. [12], obtained using the A – ψ method. This device has been one of the benchmarks for the TEAM workshops and it is interesting to note that of eleven submissions from all over the world, five were of the method 1 type and five were broadly method 2 [16].

Results are shown on Figs. 4–7.

5. Time Transient Problems

Equations such as (47), together with the appropriate magnetic scalar equation, may be expressed in matrix form as:

$$kx + cx = f \tag{48}$$

This equation can be solved using one of the classical time–stepping methods. One convenient scheme which we use [17] is the general theta method, described in ref. [18]. Some results obtained for the FELIX brick test piece, fig. (8), (also one of the TEAM workshop benchmark experiments [19]) are shown on fig. (9).

6. Moving Conductor Eddy Current Problems

Many devices, for instance, electromagnetic launchers and linear induction machines, involve conducting parts which move. The geometry of these machines is often such that full 3D computer models are required. In this section, two formulations for solving this type of problem are described; one of the methods requires the use of V inside the moving conductor and the other does not.

Only the type of moving conductor problem in which the moving member is invarient in the cross section which is normal to the direction of motion is considered here. This allows motion to be taken into account using the well known Minkowski transformation, which leads to a steady state solution for constant speed moving conductor problems. All other geometries lead to a full time transient analysis.

64

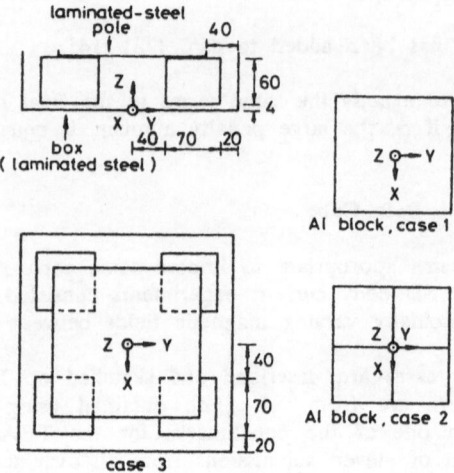

Case 1   A solid block, centrally positioned in the steel box.

Case 2   Four blocks, insulated from one another, centrally positioned.

Case 3   Four blocks, symmetrically positioned

Fig. 3   Showing the dimensions of the Bath cube experiment from reference [15]

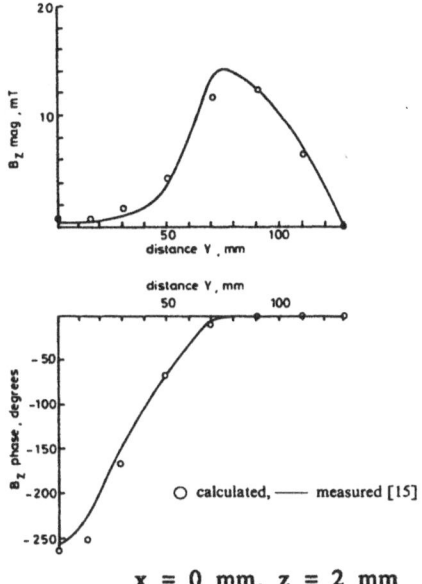

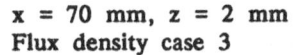

x = 0 mm, z = 2 mm

Fig. 4   Flux density case 1

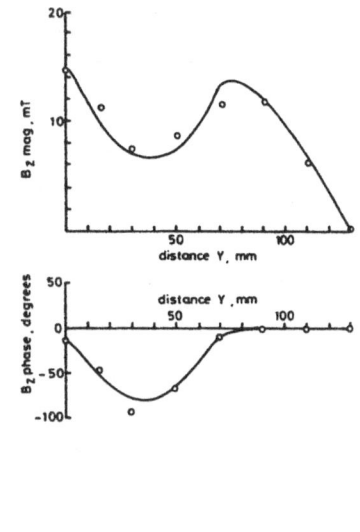

x = 50 mm, z = 2 mm

Fig. 5   Flux density case 2

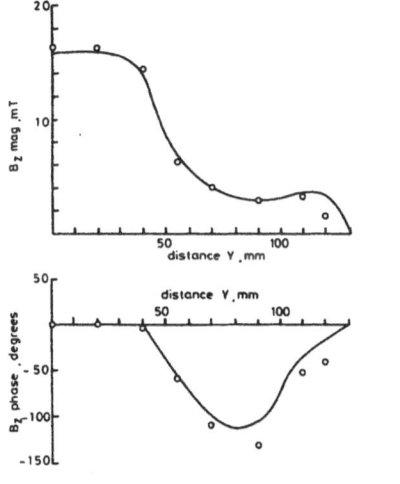

x = 70 mm, z = 2 mm

Fig. 6   Flux density case 3

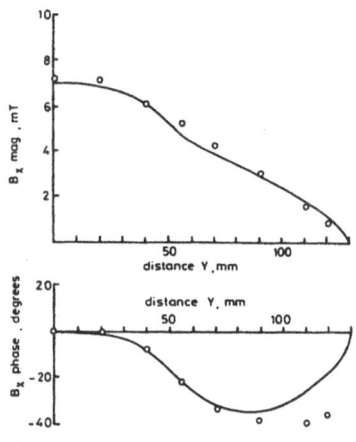

x = 120 mm, z = 20 mm

Fig. 7   Flux density case 3

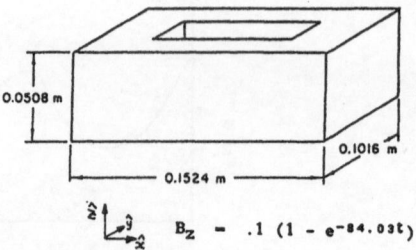

0.0508 m

0.1016 m

0.1524 m

$B_z = .1 (1 - e^{-84.03t})$

Fig. 8   The FELIX brick test piece

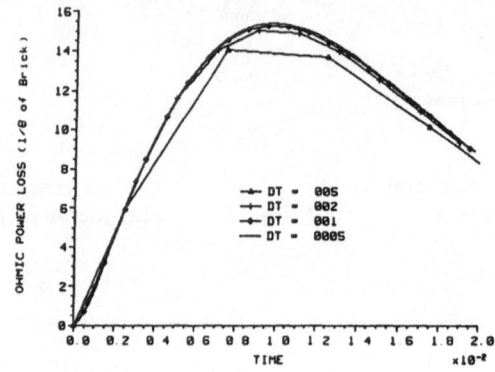

Predicted ohmic loss for various time steps
(coarse mesh)

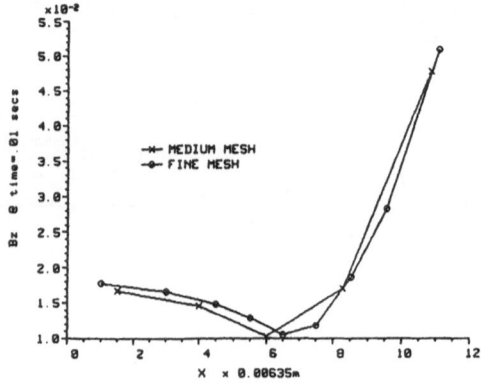

Field along x axis of FELIX brick

Fig. 9   Some calculated results for the FELIX brick

6.1 MOVING CONDUCTOR FORMULATION USING V INSIDE CONDUCTORS

In the laboratory reference frame, the moving region electric field has two components:

$$E = u \times B - grad\ V \tag{49}$$

In the above, u is the velocity of the region with respect to the laboratory and V is the electric scalar potential.

Using $B = curl\ A$, we can obtain, from equation (16):

$$curl\ \nu\ curl\ A = \sigma\ (u \times curl\ A - grad\ V)$$

and also from div J = 0:

$$div\ \sigma\ (u \times curl\ A - grad\ V) = 0 \tag{50}$$

As described in section 3.5, we gauge the problem by forcing div A = 0 throughout the conducting volume and A.n = 0 on the boundary. Eqn. (50) above applies to problems in which there are no time variations, if required, a $\partial A/\partial t$ term may be added.

The terms involving the velocity u require special treatment, upwinding.

6.1.1 *Upwinding*: When the Galerkin technique is applied to eqn. (50), large −ve terms are generated on the diagonal of the final global matrix. This typically causes oscillations in the solution and very poor results when the Peclet number, $p = \sigma h u \mu / 2.0$ is greater than 1.0 (h is the average element length in the direction of the velocity). This problem is familiar in fluid mechanics. The solution is known as upwinding.

A finite element scheme which allows different degrees of upwinding in each moving conductor element has been developed for fluid flow [20].

Usually, the integrals of eqn. (50) are evaluated using Gaussian quadrature, sampling at the normal quadrature points. Using an upwind scheme, different sampling points are used for evaluating the velocity terms only of eqn. (50) as follows:

For element e:

$$\int_{\Omega_e} N\ .\ (\sigma(u \times curl\ A))\ d\Omega \triangleq$$

$$\sum N(\epsilon)\ .\ (\sigma(u(0) \times curl\ A\ (\epsilon)))\ J(0)\ \omega$$

u(0) is the velocity evaluated at the origin of the isoparametric co−ordinates of the element, J(0) is the Jacobian of the isoparametric transformation, ω equals 8 for a 3D element and 4 for a 2D element. The location of point ϵ (this is a local co−ordinate, $-1 \leqslant \epsilon \leqslant 1$) determines the degree of upwinding.

The optimal position for ϵ has been shown to be [20]:

$$\epsilon = coth\ p - \frac{1}{p}$$

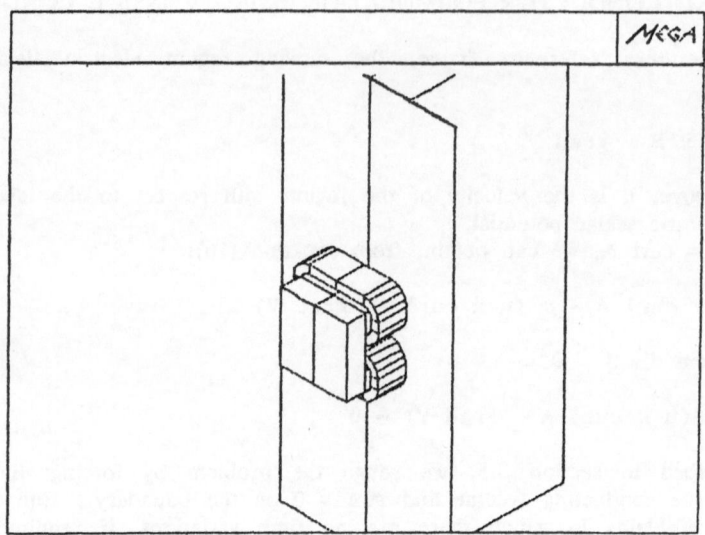

Fig. 10a   Showing a magnet moving along a conducting rail

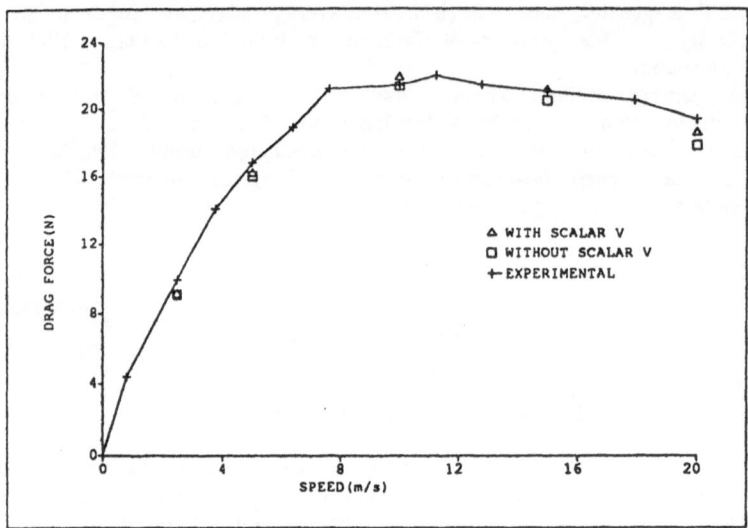

Fig. 10b   Drag force versus speed for the magnet calculated with and without the electric scalar V

6.2 MOVING CONDUCTOR FORMULATION WITHOUT V INSIDE CONDUCTORS

Starting from eqn. (50), we can make the substitution V = **A.u**.
Using **u** x curl **A** = grad (**A.u**) − (**u**.grad) **A** we can obtain

$$\text{curl } \nu \text{ curl } \mathbf{A} = -\sigma \, (\mathbf{u} \, . \, \text{grad}) \, \mathbf{A} \tag{51}$$

This equation is more economic than eqn. (50) and the results are similar [5], as shown in Fig. 10.

7. Multiply Connected Regions

It is obvious that a large saving in computer effort is possible if magnetic scalar potentials are used in non conducting regions of 3D problems, since only one variable is needed rather than three per point in space. One of the problems that arises as a consequence of using magnetic scalars occurs when conductors with holes are modelled. The regions are then multiply connected. If we take a path C, Fig. (11), through the magnetic scalar volume, passing through a hole in the conductor, we have, from Ampere's Law:

$$\oint \text{H. } d\ell = \text{total current enclosed by the path C}$$

This line integral is always zero so that the total current in the conducting loop will be zero. This may not be desirable. Several solutions to this problem have been investigated [21−23]. It is fairly obvious that if the A−V method is used inside conductors, all holes may be filled with non−conducting A regions, so that no conductor need ever be multiply connected. Other methods [21,22] involve jumps in the magnetic scalars, so that the 'hole' is spanned by a special cut through the magnetic scalar. Reference [21] describes how Lagrange multipliers may be used to join double valued scalars across a cut, and in [22] a method is described whereby the jump in the magnetic scalars is introduced as a new variable in the equation system. This option is more economic than the A−V possibility, as Fig. 12 shows [24].

8. Eddy Currents in Thin Sheets

Many practical problems involve devices incorporating thin conducting sheets. Examples arise in eddy current screening (electromagnetic compatibility), non destructive testing and induction heating. The method [25] about to be described in this section is relevant to the calculation of 3D eddy current distributions in thin plates of arbitrary shape and of small thickness compared to the skin depth of the material at the excitation frequency.

Various solutions to this particular type of problem have previously been described, among the earliest was presented in a series of papers by Carpenter et al [6,26,27].

Fields away from eddy current regions are modelled using magnetic scalar potentials and the sheet eddy currents are modelled using a surface quantity T, the curl of which is proportional to the surface current density. The field equations

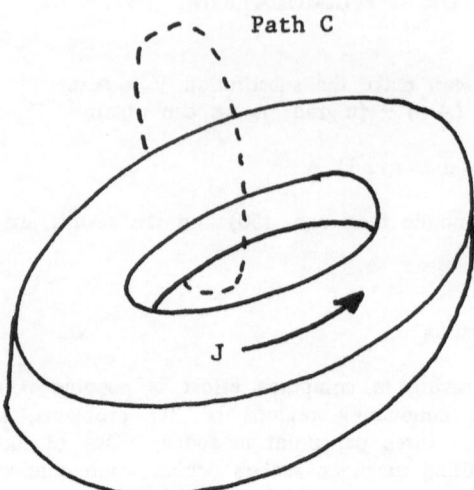

Fig. 11   Showing a conductor with one hole – a multiply connected problem

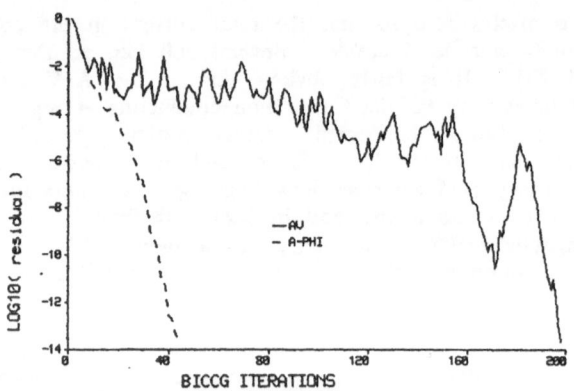

	methods	nodes	elements	equations	non-zeros	BICCG iterations
[21]	Lagrange	7504	6915	11512	301248	39
[22]	new method	7504	6915	11429	305143	41
	$AV\psi$	7504	6915	13095	450992	198

Fig. 12   Showing convergence of three methods for dealing with a multiply connected problem

were solved using either circuit analogue or finite difference techniques. Also around that time, an integral equation circuit analogue method was being developed [28].

The method described here is based on differential field equations. The regions surrounding the conducting sheets are modelled using magnetic scalar potentials and the surface current density is represented by a scalar stream function. This scheme is compatible with the A-ψ method designed to model 3D eddy currents in general thick conductors.

A thin sheet approximation method may seem superfluous given that general 3D eddy current methods are already being developed, however the thin sheet technique is more economic for this special case and, more important, very thin sheets can be modelled without introducing any of the numerical difficulties associated with volume elements that are thin compared with the other dimensions.

8.1 FIELD EQUATIONS

8.1.1 *Interface Conditions at the Conducting Sheet*: As described in section 2.2, non conducting regions are modelled using reduced or total magnetic scalar potentials. In this section we describe the sheet current field model and how this interfaces to the magnetic scalars.

If eddy currents flow in a conductor which is thin compared to the skin depth of the material, the distribution of current is essentially uniform across the thickness of the sheet.

The surface current density is usually defined as the limit of the product of sheet current density and thickness d, that is, $K = \lim Jd$, $J \to \infty$ as $d \to 0$.

If we define K as:

$$K = Jd \tag{52}$$

the sheet current can be modelled by K, clearly the thinner the sheet the more correct will be the approximation, and the more similar K will be to an ideal surface current sheet.

As K varies along the conductor surface only, K can be represented by means of a scalar quantity T existing only on the sheet surface:

$$K \propto \text{grad } T \times n \tag{53}$$

in which n is the unit normal to the sheet surface.

It will be shown that it is convenient to introduce the time derivative in the definition of T (for reasons of symmetry):

$$K = \frac{\partial}{\partial t} (\text{grad } T \times n) \tag{54}$$

Fig. 13 depicts a typical sheet conductor, completely surrounded by the total magnetic scalar potential ψ. The potentials on both sides of the sheet surface are labelled ψ_1 and ψ_2. At an arbitrary point on the sheet surface there is a jump in H x n due to the current sheet:

$$\text{grad } \psi_1 \times n - \text{grad } \psi_2 \times n = K \tag{55}$$

Continuity of B.n yields:

$$\mu_1 \frac{\partial \psi_1}{\partial n} - \mu_2 \frac{\partial \psi_{.2}}{\partial n} \tag{56}$$

μ_1 and μ_2 being the permeability on either side of the sheet.
At the edge of the sheet $\psi_1 = \psi_2$.
By substituting eqn. 52, then eqn. 6 into eqn. 54, and taking the curl we obtain:

$$\mathrm{curl} \; \frac{1}{\sigma d} \frac{\partial}{\partial t} \, (\mathrm{grad} \; T \times n) = \mathrm{curl} \; E \tag{57}$$

Using eqn. 2, an equation linking T to the component of magnetic flux density normal to the sheet may be obtained:

$$\left[\mathrm{curl} \; \frac{1}{\sigma d} \, \mathrm{grad} \; T \times n \right] . \, n = -B \, . \, n = \mu_1 \frac{\partial \psi_1}{\partial n} - \mu_2 \frac{\partial \psi_2}{\partial n} \tag{58}$$

From eqns. 55 and 54:

$$\mathrm{grad} \; \psi_1 \times n - \mathrm{grad} \; \psi_2 \times n = \frac{\partial}{\partial t} \, (\mathrm{grad} \; T \times n) \tag{59}$$

Dividing by σd and then taking the curl of eqn. 59 yields:

$$\mathrm{curl} \; \frac{1}{\sigma d} \, (\mathrm{grad} \; \psi_1 \times n - \mathrm{grad} \; \psi_2 \times n) \, . \, n$$

$$= \mathrm{curl} \left[\frac{1}{\sigma d} \frac{\partial}{\partial t} \, (\mathrm{grad} \; T \times n) \right] . \, n \tag{60}$$

Eqns. 58 and 60 define the curl of the vector $1/\sigma d$ grad $T \times n$ in terms of the derivatives of the magnetic scalar potentials. By the Helmholtz theorem, it only remains to specify the divergence of the vector and to apply suitable boundary conditions in order to define it uniquely. The divergence can be shown to be zero as follows:

$$\mathrm{div} \left[\frac{1}{\sigma d} \, \mathrm{grad} \; T \times n \right] = \frac{1}{\sigma d} \, \mathrm{div} \, (\mathrm{grad} \; T \times n)$$

$$+ \, \mathrm{grad} \; T \times n \, . \, \mathrm{grad} \left[\frac{1}{\sigma d} \right] \tag{61}$$

If we assume that σ and d are constants within each finite element, the second term on the right can be ignored, leaving the first term. Expanding this leads to: n . curl grad T − grad T . curl n, which is zero.

Fields throughout the problem region may be obtained by solving the Laplacian equation throughout the magnetic scalar volumes, together with eqns. 58 and 60 at the sheet surface.

8.2 NUMERICAL IMPLEMENTATION

As described in section 3.2, the Galerkin treatment of the magnetic scalar Laplacian leads to a surface integral of B.n (eqn. 31). At the conductor interface, this links to the B.n term in eqn. 58, as follows.

8.2.1 *Numerical Representation of eqn. 58*: Applying the same Galerkin treatment to eqn. 58 yields:

$$\int N \ curl \ \frac{1}{\sigma d} \ (grad \ T \ x \ n) \ . \ n \ d\Gamma + \int NB \ . \ n \ d\Gamma = 0 \qquad (62)$$

where N is a set of surface weighting functions.

Using the identity s curl V = curl sV − grad s x V where V is a vector and s is a scalar, yields:

$$\int \left[curl \left[N \ \frac{1}{\sigma d} \ grad \ T \ x \ n \right] \right.$$

$$\left. - \ grad \ N \ x \left[\frac{1}{\sigma d} \ grad \ T \ x \ n \right] \right] \ . \ n \ d\Gamma + \int NB \ . \ n \ d\Gamma = 0 \qquad (63)$$

Applying Stoke's theorem to the first term in eqn. 63 leads finally to:

$$\oint N \ \frac{1}{\sigma d} \ grad \ T \ x \ n \ . \ dl$$

$$- \int \left[grad \ N \ x \left[\frac{1}{\sigma d} \ grad \ T \ x \ n \right] \right] \ . \ n \ d\Gamma + \int NB \ . \ n \ d\Gamma = 0 \qquad (64)$$

The line integral (first term in eqn. 64) represents the tangential component of E around the edge of the sheet.

8.2.2 *Numerical Representation of eqn. 60*: Applying Galerkin to eqn. 60 as before:

$$\int N \ curl \ \frac{1}{\sigma d} \ (grad \ \psi_1 \ x \ n - grad \ \psi_2 \ x \ n) \ . \ n \ d\Gamma$$

$$- \int N \ \text{curl} \left[\frac{1}{\sigma d} \frac{\partial}{\partial t} (\text{grad } T \times n) \right] . \ n \ d\Gamma = 0 \tag{65}$$

This leads to:

$$\oint N \frac{1}{\sigma d} (\text{grad } \psi_1 \times n - \text{grad } \psi_2 \times n) . \ dl$$

$$- \int \left[\text{grad } N \times \frac{1}{\sigma d} (\text{grad } \psi_1 \times n - \text{grad } \psi_2 \times n) \right] . \ n \ d\Gamma$$

$$- \oint N \frac{1}{\sigma d} \left[\frac{\partial}{\partial t} (\text{grad } T \times n) \right] . \ dl$$

$$+ \int \text{grad } N \times \frac{1}{\sigma d} \left[\frac{\partial}{\partial t} (\text{grad } T \times n) \right] . \ nd\Gamma = 0 \tag{66}$$

8.2.3 *Interface Conditions and Continuity*: The magnetic scalar volumes are discretised according to the usual finite element procedure. Expressions such as eqn. 31 are assembled for each finite element. Both total and reduced scalar potentials are treated in this manner.

It will be noted that inside magnetic scalar volumes the surface terms in eqn. 31 ensure continuity of B.n. On the sheet conductor surfaces, three variables exist. The surface integrals of eqn. 31 that interface the conducting sheet are used to link ψ with T on the surface, i.e. from eqns. 31 and 64 for a 'ψ_1' node of Fig. 13:

$$\int \mu_1 \text{ grad } N . \text{ grad } \psi_1 \ d\Omega$$

$$+ \int \left[\text{grad } N \frac{1}{\sigma d} \text{grad } T \times n \right] . \ n \ d\Gamma = 0 \tag{67}$$

A similar equation in the other 'ψ_2' node may be obtained in the same way, so that eqn. 56 is satisfied. A simultaneous solution of equations similar to eqns. 66 and 67 is carried out. The terms involving the line integrals of eqns. 64 and 66 cancel throughout the interior of a sheet to yield continuity of the tangential component of E, allowing for possible jumps in σd at element edges.

If T is zero at the edge of a sheet, the condition that no current flows out of the sheet edge is obtained. Zero E tangential to a sheet edge is often required in certain symmetry situations, this is the natural boundary condition for eqns. 64 and 66.

It will be noted from eqn. 54 that simple continuity of T between adjacent sheet elements ensures that the component of J normal to that common edge is continuous.

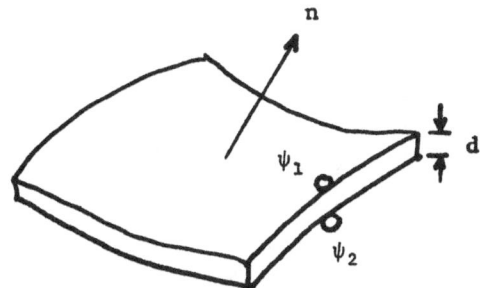

Fig. 13   Showing a typical thin conducting sheet

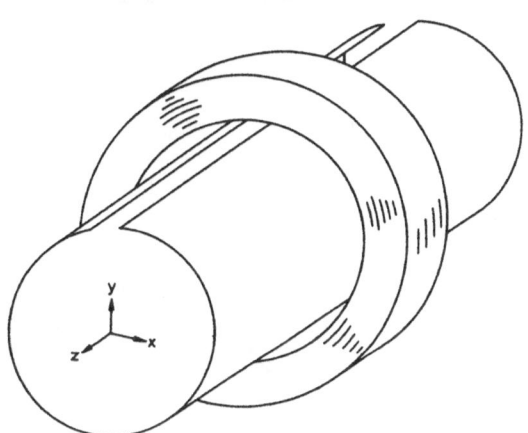

Aluminium cylinder   Coil

Length	=	305 mm	
Inside diameter	=	100 mm	
Thickness	=	0.88 mm	
Gap	=	14.5 mm	
Conductivity	=	$3.278 \times 10^{7}\Omega^{-1}m^{-1}$	

Coil:
Inside diameter = 110 mm
Outside diameter = 170 mm
Width = 65 mm
163 A turns

z equals zero at the cylinder centre

Fig. 14   A split cylinder test problem

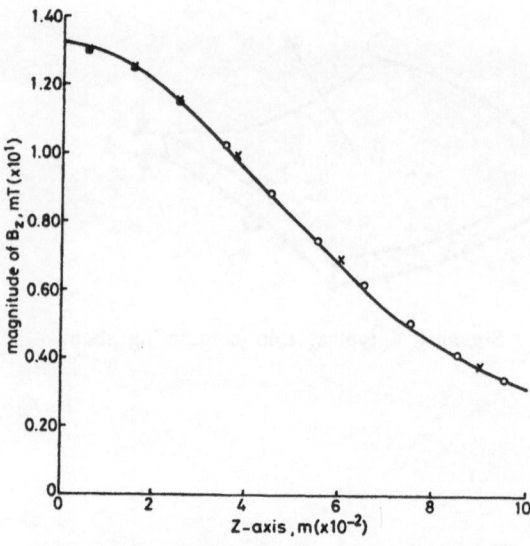

Magnitude of flux density along the z axis

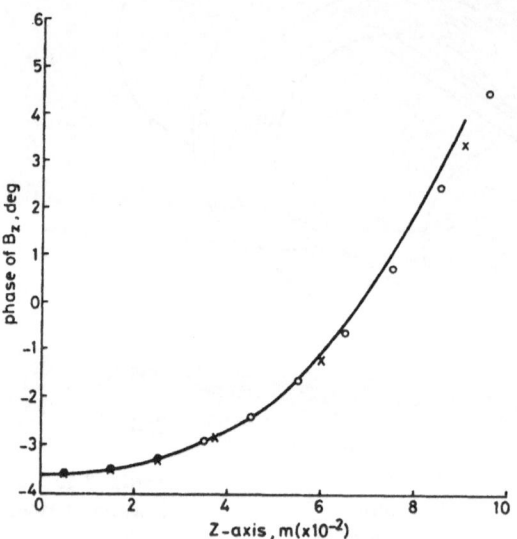

Phase of flux density along the z axis

Fig. 15 Showing calculated and experimental results for flux density along the axis of the split cylinder test problem

The final set of equations is symmetric, and may be solved as usual by means of the preconditioned conjugate gradient technique.

8.3 EXAMPLE CALCULATIONS USING THE THIN SHEET METHOD

A thin walled split cylinder test problem is shown in fig. 14. Eddy currents are excited in the shell by the surrounding coil carrying 50 Hz current. The cylinder was modelled using the MEGA 2 and 3D electromagnetics package, using $A - \psi - \varphi$ and also $T - \psi - \varphi$. Graphs of magnitude and phase of flux density along the z axis are shown on Fig. 15. The results are similar for this problem where the sheet thickness is 0.88 mm. However, induction heating problems with sheets of 30 x 10^{-6} m thick have also been solved with the T method. These would be uneconomic using volume elements.

9. Conclusions

3D electromagnetic field calculation has advanced greatly in the past decade. This progress has been largely driven by the increase in available computer power. Many different techniques have been developed for 3D electromagnetics modelling. It appears at present that no single technique will be useful for all general problems, there is no single formulation which dominates the 3D field in the same way as two dimensional calculations are dominated by a single component of magnetic vector potential. It is the author's opinion that the 3D software packages of the future will have to be capable of modelling problems using several formulations.

10. References

1. Last three COMPUMAG conference proceedings:
   Tokyo IEEE Trans Mag, Vol. 26, No. 2
   Graz IEEE Trans Mag, Vol. 24, No. 1
   Fort Collins IEEE Trans Mag, Vol. 21, No. 6

2. Last three INTERMAG conference proceedings:
   Brighton IEEE Trans Mag, Vol. 26, No. 5
   Washington IEEE Trans Mag, Vol. 25, No. 5
   Vancouver IEEE Trans Mag, Vol. 24, No. 6

3. Zienkiewicz, O.C., Lyness, J. and Owen, D.R.J: 'Three dimensinal magnetic field determination using a scalar potential – a finite element solution', IEEE Trans., 1977, MAG–13, pp. 1649–1656.

4. Simkin, J. and Trowbridge, C.W.: 'On the use of total scalar potential in the numerical solution of field problems in electromagnetics', Int. J. Numer. Methods Eng., 1979, 14, pp. 423–440.

5. Rodger, D., Leonard, P.J. and Karaguler, T.: 'An optimal formulation for 3D moving conductor eddy current problems with smooth rotors', to be published, IEEE Trans Mag, Vol. 26, No. 5.

6. Carpenter, C.J.: 'Comparison of alternative formulations of 3–dimensional magnetic–field and eddy–current problems at power frequencies', Proc. IEE, 1977, 124, (11), pp. 1026–1034.

7. Rodger, D., and Eastham, J.F.: 'Finite element solution of 3D eddy current flow in magnetically linear conductors at power frequencies', IEEE Trans, 1982, MAG–18, pp. 481–485.

8. Zienkiewicz, O.C.: 'The finite element method' (McGraw–Hill, 1977, 3rd Edition).

9. Nedelec, J.C.: 'Mixed elements in R^3', Numer. Math., 1980, 35, pp. 315–341.

10. Bryant, C.F., Emson, C.R.I. and Trowbridge, C.W.: 'A comparison of Lorenz gauge formulations in eddy current calculations', IEEE Trans Mag, Vol. 26, No. 5., pp. 430–433.

11. Albanese, R. and Rubinacci, G.: 'Formulation of the eddy–current problem', Proc IEE, Vol. 137, Pt. A, No. 1, Jan 1990, pp. 16–22.

12. Rodger, D.: 'A finite element method for calculating power frequency three dimensional electromagnetic field distributions', Proc IEE, Vol. 130, Pt. A, No. 5, 1983, pp. 233–238.

13. Rodger, D., Karaguler, T., Leonard, P.J.: 'A formulation for 3D moving conductor eddy current problems', IEEE Trans Mag, Vol. 25, No. 5, Sep 1989, pp. 4147–4149.

14. Renhard, W., Stogner, H. and Preis, K.: 'Calculation of 3D eddy current problems by the finite element method using either an electric or a magnetic vector potential', IEEE Trans Mag, Vol. 24, No. 1, Jan 1988, pp. 122–125.

15. Davidson, J.A.M. and Balchin, M.J.: 'Experimental verification of network method for calculating flux and eddy–current distributions in three dimensions', IEE Proc. A, 1981, 128, (7), pp. 492–496.

16. Turner, L. (editor): 'Benchmark problems for the validation of eddy current computer codes', COMPEL, Vol. 9 (1990), No. 3.

17. Leonard, P.J. and Rodger, D.: 'Some aspects of 2 and 3D transient eddy current modelling using finite elements and single–step time marching algorithms', Proc. IEE, Pt. A, Vol. 135, No. 3, March 1988, pp. 159–166.

18. Zienkiewicz, O.C., Wood, W.L. and Hines, N.W.: 'A unified set of single step algorithms', IJNME, Vol. 20, pp. 1529–1552, 1984.

19. Turner, L. (editor): 'Papers on benchmark problems for the validation of eddy current computer codes', COMPEL, Vol. 7, March/June 1988.

20. Hughes, T.J.R.: 'A simple scheme for developing 'upwind' finite elements', IJNME, Vol. 12, pp. 1359–1365.

21. Rodger, D. and Eastham, J.F.: 'Multiply connected regions in the A − ψ three dimensional eddy current formulation', Proc. IEE, Pt. A, Vol. 134, No. 1, Jan 1987, pp. 58–66.

22. Leonard, P.J., Rodger, D.: 'A new method for cutting the magnetic scalar potential in multiply connected eddy current problems', IEEE Trans Mag, Vol. 25, No. 5, Sep 1989, pp. 4132–4134.

23. Biro, O. and Preis, K.: 'Finite element analysis of 3D eddy currents', IEEE Trans Mag, Vol. 26, No. 2, pp. 418–423.

24. Leonard, P.J. and Rodger, D.: 'Comparison of methods for cutting the magnetic scalar potential in eddy current problems with holes', COMPEL, Vol. 9A, ISSN 0332–1649, pp. 55–58.

25. Rodger, D. and Atkinson, N.: 'A finite element method for 3D eddy current flow in thin conducting sheets', Proc. IEE, Pt. A, Vol. 135, No. 6, July 1988, pp. 369–374.

26. Carpenter, C.J. and Djurovic, M.: 'Three–dimensional numerical solution of eddy currents in thin plates', Proc. IEE, 1975, 122, (6), pp. 681–688.

27. Carpenter, C.J. and Wyatt, E.A.: 'Efficiency of numerical techniques for computing eddy currents in two and three dimensions'. Conference on computation of magnetic fields, Oxford, 1976, pp. 242–250.

28. Turner, L.R.: 'An integral equation approach to eddy–current calculatins', IEEE Trans Mag, 1977, MAG–13, (5), pp. 1119–1121.

CLASSIFICATION OF SOFTWARE AVAILABLE AND CRITERIA OF APPROACH

GIORGIO MOLINARI
Dipartimento di Ingegneria Elettrica
Universita' di Genova
11a, Via Opera Pia
I-16145 Genova Italy

ABSTRACT. In this paper it will be presented an overview and a classification of the software available for industrial applications in the area of Computer Aided Electromagnetic Analysis, and criteria will be suggested to select and evaluate it according to the user requirements.

1. Introduction

It is assumed in this paper that the reader has already become acquainted with the basic aspects of Electromagnetic Computer Aided Analysis, or Computational Electromagnetics, and is therefore, at least at the level of definitions, familiar with terms and concepts such as discretization of electromagnetic partial differential equations, both with integral and differential approaches, and is knowledgeable in principle of some of the various possible methods of solution, such as Boundary Element and Finite Element, to name some of the most common ones.

Some awareness is also assumed of the general structure of the computer codes implementing the above mentioned methods, usually composed of a main code and interactive pre- and post-processors, and of their requirements in terms of computer resources, at present best met usually by products in the class of workstations.

Starting from this background, the paper will attempt to provide an overview and a classification of the software of interest for industrial applications in the area, also recalling briefly the historical developments and the main reasons that have shaped the current situation.

On the basis of this assessment, the main criteria to be taken into account for a correct evaluation of the matching between code features and user requirements will be then outlined in some detail, stressing the points that have proven of major importance in the experience of the author.

Finally, a brief survey of the market situation will be provided, listing the Firms known to the author as Vendors of software for electromagnetic analysis, providing some information of their features, and offering some suggestion for the definition of a sensible policy in the area.

Y. R. Crutzen et al. (eds.), Industrial Application of Electromagnetic Computer Codes, 81–100.
© *1990 ECSC, EEC, EAEC, Brussels and Luxembourg.*

2. Brief Historical Overview

In order to provide a more detailed understanding of the present situation of Electromagnetic Computer Aided Analysis, it is useful to describe briefly some main steps of its historical development.

An initial consideration that is to be made to put the issue in the right perspective is that the development of Electromagnetic Computer Aided Analysis started later and at a lower pace with respect to other scientific areas, such as Thermal and Structural Analysis. Even if it is certainly not easy to define fully the motivations of this situation, a few reasons that have certainly contributed to this delay, and that can be useful to understand better the subsequent pattern of development, can be easily given.

A first important, albeit obvious, consideration is that the dimension of the market for Electromagnetic Analysis tools, even if expanding, is certainly smaller than that of Thermal and particularly of Structural Analysis, and is then potentially less profitable to cover.

A second important point is that the design of Electromagnetic devices and systems has been traditionally based on a circuit approach, significantly simpler to master and use with respect to the field approach, that has proven in the past accurate enough for practical industrial applications, at least in most cases, reducing in the initial years the drive to the development of Electromagnetic Analysis tools.

Finally, Electromagnetic Analysis, with respect to Structural Analysis, presents substantial difficulties even at an entry level, because of the variety of materials involved, with the attendant complexity of governing equations, the complicate geometric shapes required, the frequent presence of nonlinear and transient effects, and the need to handle open boundaries for a significant range of problems.

The above reasons, besides determining the delay of development previously mentioned, have also shaped a very uneven initial interest in the area, that has started essentially for advanced designs, where the approximations provided by the standard circuit approach proved to be inadequate. This has happened initially for high energy physics studies, that have fostered in the sixties, in a number of research Laboratories, a significant development of these techniques.

The methods used at this stage, that could be defined as the initial one for the Electromagnetic Computer Aided Analysis, were based mostly on the superposition of analytical solutions, and produced codes like COILS, MAFCO and its several variants, and others, bound of course to linear problems. A significant development in these years was the implementation of TRIM, using a finite difference approach on triangular meshes, and then able to handle in principle also nonlinear problems. The meshing principle of TRIM, even if somewhat cumbersome by today standards, is still in use, mostly in research centres on physics around the world, in one of the many spring off of TRIM that have developed over the years in a variety of sites, generally named POISSON, PANDIRA or SUPERFISH.

Generally somewhat later with respect to the activity in the area of high energy physics, an interest for these techniques started to arise also in industrial environments, that however, being in general much more cost-conscious than research laboratories, limited initially the activity to the areas more likely to produce a rapid return of the investment.

This meant that the interest arose essentially in very large industrial organizations and in a limited number of University, and was initially mostly confined to the analysis of large, expensive and design-critical machines and devices. In turn, this fact generated an initial difficulty to circulate in the scientific community the obtained results, since achievements obtained by industrial organizations were confined mostly to internal use.

Typical of this period, that covers roughly the seventies and that we could call an intermediate stage of development, is the implementation of the code MAGGY within Philips and of a range of codes within General Electric in the U.S.. In the same period development was of course also progressing for high energy physics applications, particularly with the development of POISSON already mentioned, and with the introduction of GFUN, a 3D code based on an integral approach, developed at the Rutherford Appleton Laboratory in the U.K., and then distributed to other research centres in physics.

In the second part of the seventies, the interest started to spread out of the initial groups of developers, interesting a larger number of Universities and Research Centres and starting to cover a wider range of applications. Very important to this end, in the opinion of the author, was the launching of the series of Conferences COMPUMAG, specifically devoted to the discussion of methods and results in the numerical evaluation of electric and magnetic fields, that allowed to gather universities, research laboratories and advanced industrial development teams working in the field.

This trend continued and strengthened in the eighties, producing a situation of research and development activity more evenly distributed among universities, specialized scientific software house, research Centres and industrial development Laboratories.

The present situation, in the author's experience, is characterized by a rather strong and established scientific development activity, debating all aspects of numerical field computation of electromagnetic fields, in the previously mentioned series of Conferences COMPUMAG and in several other scientific events that have flourished to cover the field.

However, for the reasons explained above, that find a confirmation in the lively scientific activity on the subject, the area has not reached yet a complete and uniform coverage of all industrially significant modelling situations, evenif the market offer already a rather wide and diversified offer of commercial codes. The codes offered present a rather significant spread of features as a function of the specific application areas best covered by the code developers and are still significantly evolving to take on board the most recent scientific developments, particularly in 3D.

A further note that should be added to complete the picture is the recent trend of several structural and thermal analysis code developers to start to offer in their environment also modules covering aspects of electromagnetic analysis, providing an additional spread of the market offer.

In the following sections we will try to indicate possible ways to classify the software available and to define criteria for a correct selection with reference to the user needs, in order to provide to perspective users a guidance in the selection of their policy in electromagnetic analysis.

3. Classification of Computer Aided Analysis Software

In the opinion of the author, the first key decision to be taken to prepare the ground for the selection of a strategy is relevant to the "width of geometric coverage" of the electromagnetic analysis activity required, that will then allow the selection of possible policies in the area; it is then useful to have a clear idea of the various existing categories of electromagnetic analysis software from this standpoint.

Even if it is obviously difficult to provide clear-cut classifications in this as in most areas featuring a wide variety of products, it is possible in this case, in the author's experience, define three main categories of codes or packages in which most, if not all, computer aided analysis codes and packages could roughly fit as far as "width of geometric coverage" is concerned.

In the following subsections the three categories will be defined and explained, trying to outline as well general advantages and disadvantages, the type of applications best suited to their features and the trend of industrial applications of each of them.

3.1 "PROBLEM-SPECIFIC" CODES AND PROGRAMS

This category covers the software devised to face electromagnetic analysis problems for a single device, or for a limited class of devices with very significant functional and geometrical similarities.

The obvious advantage of this approach is that the programming effort can be kept minimal, taking advantage of the constraints on the allowed shapes to simplify geometric definitions and meshing, and producing then codes generally of limited size, provided with simple or relatively simple commands, and often fast or relatively fast to execute.

Because of the relative simplicity of the environment, this category of codes is usually also the first in which complex equations or advanced modelling of materials are implemented.

The disadvantages stem out of the same root, and are relevant to the extreme stiffness of the codes, that cannot generally be used for the solution of problems different from the original one, providing a very specific coverage of analysis problem.

Difficulties can also arise for managing modifications or extensions of coverage, since this code are usually not provided with powerful and flexible service routines and features, so that any change requires modifications to the original code and then a rather advanced knowledge on the code implementation.

In an industrial environment, the most profitable use is generally done for frequent and important computations of design features of a constant type, and then unaffected by the code stiffness, or for the solution of seldom used problems of single parameter selection.

This category of codes and program is of course of greater and continuing importance for scientific purposes and as a test bed for advanced development, since the simplicity of the environment allows ease of modification and test of innovative algorithms.

Because of their narrow target, for an industrial acquisition these codes cannot be generally found in the software market "off-the-shelf", and should be either specifically commissioned to research centres or advanced software houses, or developed within the perspective end user's firm, when the required expertise is available.

However, this implies a relatively long time of acquisition and generally a relatively high cost, since the development effort should be covered by a single user, so that, in the experience of the author, this approach is at present shrinking in overall industrial coverage, particularly if internal development is required.

3.2 "DEVICE-ORIENTED" CODES

This category of codes covers a larger area with respect to the previous one, taking care of a whole class of devices, such as, say, a series of squirrel cage induction motors. This implies that the ability to define a wider class of geometries is required, that however all share a basic similarity of features.

The basic advantages are similar to those of the previous category, even if with some less emphasis since the ability to master a wider class of devices generally affects somewhat simplicity of use, size of the code and execution times, that however are usually manageable and practical.

Disadvantages are connected partly to the limitations mentioned for the previous class, again somewhat relaxed however as far as stiffness and absence of user features are concerned; codes of this type can certainly cover design evolution and tuning, but are generally unable to face significant design changes.

Further problems can be determined by the need to incorporate into the code proprietary company information, that usually prevent the possibility to acquire the software externally and forces to resort to internal development, with the consequent inability to share the cost with other users.

As a consequence of the above features, the use of "device-oriented" codes in an industrial environment is generally adopted only as a support tool for the design of "mature technology" devices, guaranteeing the required stability of main design features. Furthermore, because of the relatively high development cost and the inability to share it with other users, this becomes generally feasible only for devices to be mass produced by large industrial firms, usually provided with internal software development departments.

Even for such a limited "niche" market, the trend for industrial applications is downward, in the experience of the author, since codes in this category are increasingly challenged by those described in the next subsection.

3.3 "GENERAL PURPOSE" CODES AND PACKAGES

In this category fall the codes that, in principle, pose no "a priori" limitation to the class of geometry they are able to handle. The name of the category is placed in inverted commas, however, since this does not mean that any code in this category will be able to handle any geometric shape.

In fact, a major division in sub categories is necessary, to distinguish codes handling plane and axisymmetric geometries (2D codes) from those able to handle three-dimensional geometries without symmetries (fully 3D codes), and also codes covering an intermediate situation, broadly defined "2 1/2 D", can be found.

However, once the basic subcategory has been clearly defined, codes of this type should be able to handle any geometry of the proper type defined by the user, at least until the complexity and/or the difference in dimensions of the geometry become such to imperil the ability of the code to produce an acceptable mesh or to provide the required precision.

This implies a very significant effort to code a robust and reliable environment for the definition and meshing of the geometry, particularly in 3D, and generally requires a graphic interactive approach (usually performed by a separate ancillary code, named pre-processor) to define the geometry, control the consistency of results and provide corrections when required.

A similar effort is then required to analyze results, that also require a graphic interactive environment, provided with a significant set of post-viewing and post-processing features since the results of interest are a function of the specific device and cannot then be clearly defined in advance. Also these functions are usually performed by a separate interactive program, named post-viewer or post-processor.

The resulting advantages of this approach are a good to very good coverage of geometric shapes and classes of devices, the availability of sets of powerful to very powerful (and sometimes also user friendly) commands, both for pre- and post-processing, and increasingly good possibilities to face difficult problems, at present particularly in the less demanding 2D environment.

The disadvantages, as can be imagined, are the large to very large size of the codes, with the consequent need of significant computer resources and execution times, that range from significant to very demanding, particularly for nonstatic 3D applications.

A further costly feature is the need for a significant training of the user, again particularly in 3D, since the powerful commands also generally mean a complex environment, not easy to use and master fully, at least at the present stage of development.

Because of its flexibility, this category of codes is well suited to cover a wide range of applications, with the possible exception of those optimally suited to "problem-specific" codes described in subsection 3.1.

The large investment required by the pre- and post-processors usually makes an internal development unfeasible even for large industrial organizations, unless a commercial exploitation is envisaged; in fact, the range of commercial offers is very significant and widening, and allow already, even if the coverage is not complete yet as previously mentioned, an adequate solution for a wide range of industrial applications and even customization to specific needs.

The trend is then that of a rapid expansion at decreasing costs, since the widening customer base allow a larger subdivision of development costs, and the general geometric coverage, at least in principle, allows "off-the-shelf" delivery and then shorter times to full exploitation, even if training, as previously mentioned, is likely to be significant.

4. Criteria for a Correct Selection

On the basis of the classification provided in the previous Section, a first key decision on the type of software required for the specific needs of each industrial application as far as "width of geometric coverage" is concerned can now be made. A direct consequence of this decision, because of the features of the various categories previously described, is the choice between internal or full custom development on one side and acquisition of standard or customized commercial software on the other.

In fact, if the optimal type of software defined falls in the first two categories of the previous Section, internal or full custom development is likely to be a forced choice; in this case little can be said in general as criteria of selection, since the specific requirements are likely to be extremely problem-dependent, and the development should anyway be performed or closely followed by personnel provided with a significant expertise of design requirements teaming up with electromagnetic analysis experts to implement and validate the software.

Because of the overall features of electromagnetic analysis market briefly sketched in the previous Section however, this is not likely to be a frequent situation. In most cases it is likely to appear much more cost-effective and viable the acquisition of some kind of commercially available "general purpose" code or package, with or without customization and assistance by the development site or by a research centre according to the type of analysis problems to be faced.

Since this market, as previously outlined, is at present in a situation of rather rapid development and presents a wide range of offer with vastly different features, it appears then important to screen in some detail the most important issues to examine before the acquisition, to be able to perform a correct selection of the code best suited to the specific needs of a given industrial application.

When selecting a large and expensive industrial equipment, a number of issues are to be closely examined, and usually are, to make sure that the product is able to provide a return of investment matching the direct and indirect costs related to its acquisition.

Large software codes and packages are no exception to this rule, even if it could appear strange to examine in an industrial light an apparently immaterial product; however, costs and potential benefits are far from immaterial, so that software products deserve the same careful examination of an automatic lathe or a flexible manufacturing centre.

When analyzing the possible acquisition of an electromagnetic analysis code or package, a significant documentation effort should be started, collecting on the products under examination the information required for a correct screening of features.

In the experience of the author, the most important information to be collected could be grouped in the five following categories:

1 – Coverage of electromagnetic problems

2 – User features

3 – Hardware required

4 – Technical support

5 – Cost structure

The most important features of the information to collect in each of these category will be reviewed in some detail in the following subsections.

4.1 COVERAGE OF ELECTROMAGNETIC PROBLEMS

At present, as already outlined in the previous Sections, electromagnetic analysis codes, because of the complexity of the modelling involved, do not cover yet the whole range of problems of industrial interest, so that a clear definition of the user requirements becomes essential, and should be matched as closely as possible with the capability of the code selected.

A first important selection to be made is that among codes having 2D and 3D capabilities, as defined in the previous Section. The type of geometry handled in this respect affects significantly the size, complexity and cost of the analysis code, and is consequently important for a proper selection.

At present, since 3D codes are very large and computer resource intensive, particularly to model dynamic and transient phenomena, they are usually employed only when the two-dimensional cases, much simpler and faster to run, have been fully explored.

This suggests to have available in general codes able to handle both 2D and 3D problems, to have a complete coverage and optimal operating conditions. However, the specific features of the industrial application could suggest other solutions, whose operational boundary should be anyway defined as clearly as possible in advance.

Once this important initial selection has been made, attention should be paid to the coverage provided by the codes under examination of the various subclasses in which electromagnetic analysis problems could be divided.

From this standpoint, most codes presently on the market can be grouped in one or more of the subclasses outlined below, each of which will be briefly described and commented in the following.

4.1.1 *Electrostatic and Magnetostatic Codes.* Generally offered together, these codes cover the areas simpler to implement and providing the faster solutions.

However, electrostatic and magnetostatic problems each present specific features whose presence is not necessarily granted, and is generally a function of the previous application background of the code developers.

Examples of such features are the ability to handle floating conductors (that is, structures whose potential is constant but not assigned), that could be of interest for electrostatic applications and is useless in magnetic problems, and the ability to handle magnetic-type nonlinearities, having opposite interest. A detailed understanding should then be gained of the precise coverage of the code version under examination, and of the extensions under implementation or planned.

Codes with these capabilities are generally offered alone for 3D geometries, whereas they are generally bundled with additional coverages in 2D.

4.1.2 *Sinusoidal Steady-State Eddy Currents Codes.* This kind of code, intrinsically limited to linear or linearized problems, allow to obtain a solution of a dynamic problem with a single solution, using a complex variable. Because of this feature, they allow to minimize computer resources, and are usually the first coverage of dynamic problems offered by 3D codes.

In 2D, this type of capability is generally bundled with static ones, and frequently also with those described in the next paragraph.

4.1.3 *General Transient Eddy Current Codes.* The solution of this type of problems requires time discretization, and multiple unknowns per node in 3D. As a consequence, these problems are anyway computationally intensive, and extremely so in complicate 3D geometries and with nonlinear materials.

Often this type of capabilities is offered together with that of the previous paragraphs in 2D, whereas in 3D they can be either alone or grouped with the previous coverages.

4.1.4 *"Advanced Modelling" Codes.* At present, the three previous paragraphs describe what could be called "standard" capabilities, even if in 3D only a few environments offer at present a complete coverage.

However, particularly in 2D, it is possible to find in specific codes some additional coverages, that could be of significant interest for a given industrial applications.

The most common ones are modelling of anisotropic, inhomogeneous or histeretic materials, ability to solve voltage driven problems or to handle moving components, coverage of limited skin-depth problems, capability to define particle trajectories and some coverage of coupled problems, particularly thermo-magnetic couplings.

However, even if not implemented yet in the current version, these features could frequently be added to existing codes, and the addition is frequently driven by customer needs. Interested users should then, when a feature of interest is not present, ask for possibility, time and cost of extension to coede developers.

4.1.5 *"General Formulation" Codes.* This class of codes, mentioned here for completeness, allows to solve in principle a wide class of partial differential equations, usually at least elliptic. They should then, in principle, be able to solve many problems in electromagnetic analysis as well, providing a significantly wider coverage than usual.

However, there are very few offers of this kind on the market, and they have not, in the experience of the author, obtained so far a very significant commercial success. This is particularly true for industrial applications, because of the additional difficulty of providing the most practical user features for the whole range of analysis areas covered in principle.

4.2 USER FEATURES

As "user features" it is intended here the complex of the pre- and post-processing commands and capabilities made available to the user on the type and models of computers and peripherals supported by the code.

In the experience of the author, these features are extremely important for an efficient exploitation of the code, particularly in an industrial environment.

They should then be evaluated carefully by the potential user, backed by experts if necessary, in an environment as close as possible to that of the future industrial use.

The best way to perform this type of evaluation is to perform benchmarks on problems of real interest to the user, performing "hands-on" solutions with the assistance of technical experts of the Vendor, to test the code features in problems as close as possible to the real ones.

It is of course very difficult to provide a comprehensive list of the features of importance, since each of them is bound to have a very different interest for different applications; however, it is possible to give here a short list of some of the items usually of greater importance, grouped into four main categories.

Data definition: completeness and ease of use of the set of commands available to define geometries, availability of material libraries and handling of additions, coverage of boundary and interface conditions, availability and ease of use of parametric definitions of geometries.
Grid generation: Availability of commands for manual, semiautomatic and/or automatic meshing, possibility of definition and use of "superelements" when applicable, treatment of open boundaries.
Post-viewing: Support of graphic peripherals, handling of contouring routines, availability of colour or 3D plots of the solution, presentation of values of the solution on lines or surfaces, windowing facilities.
Post-processing: Computation of derivatives, components and absolute value of fields, availability of routine for the computation of force, torque, inductance and capacitance, handling of line and surface integrals for the evaluation of integral quantities in general, possibility to compute

user-defined functions.

4.3 HARDWARE REQUIRED

Electromagnetic analysis code, just like analysis codes in other disciplines, are necessarily large, and even more so if they are provided with good user features and multiple analysis capabilities.

A well developed code of this type is seldom smaller then 20,000 statements and a complete package of codes providing a good general coverage can well reach several times this figure; this in itself means that portability to other computers and operating systems is far from obvious and necessarily successful. Further constraints are then added by the pre- and post-processing environments, whose graphic requirements should be matched with the graphic peripherals available.

Since, besides having a large number of instructions, these codes also generally require large amounts of main memory and disc space and long execution times, particularly in 3D, they are generally supported by the manufacturers on some classes of computers only, and, within each class, only on some models of some manufacturers.

It appears then useful to review very briefly the suitability of the main classes of computers to this type of codes, as defined by the average support provided by manufacturers to computers of each class.

Workstations. Widest choice of offers. Excellent structural matching between application and computer architecture, particularly for the upper range, that hosts already most of the full-fledged 3D codes. Potential for further improvements with the recent announcement of superworkstations with vector and superscalar features.

Minicomputers. Very wide choice of offers. Even if at present suffering, for Computer Aided Analysis applications, the competition of recently introduced workstations, they are the native development environment of most commercial codes, and are then still significantly supported, particularly for the models most diffused in technical applications.

Personal Computers. Rather wide choice of offers for entry level 2D analysis codes. However, the intrinsic limitations of the MS/DOS operating system, in spite of the constantly increasing computational power, make this environment unsuited and generally discouraged for full fledged 3D applications or even for 2D transient cases. The situation can differ for top-of-the-line PC running a Unix operating systems, in which case however both in price and performance they should just as well be reclassified as workstations.

Supercomputers. Some offers available for 3D codes, for very large and computer-intensive applications. Expansion of coverage threatened however by development of superworkstations.

Mainframes. Rather limited range of offers, usually covering the larger and most mature codes. Even if large organizations tend at present to continue to ask for support in this environment, it does not appear particularly suited to CAD applications, and its use for new installations does not appear competitive and is then discouraged.

On the basis of the above guidelines, the user should then select the hardware operational environment, with reference to its requirements and to the models and manufacturers supported by the code Vendor.

In the experience of the author, it is absolutely advisable to install the code on a combination of computer and peripherals explicitly supported by the code vendor, particularly for use in an industrial environment.

Other solutions, even if they could appear less expensive, usually provide difficulties of installation, unusual run time errors, difficult to trace back and fix, and inability by the software Vendor to provide a reliable maintenance and support. The only viable alternative is to obtain by the code Vendor the extension of coverage and continued support, of course covering the extra cost required.

4.4 TECHNICAL SUPPORT

The importance of maintenance and continued support of the code has been implicitly stated already in the previous subsection, but it deserve a more explicit stressing and coverage.

In fact, when a computer code or package is larger then ten thousands statements, the probability that it does not contain inconsistencies or errors tends to become negligible.

Particularly for an industrial use of the code, this makes extremely important to have available a reliable and timely support of the code or package, that could provide advice and suggestions on how to solve run time "traps" on specific applications, and implement then fixes to the "bugs" in further releases of the code.

This kind of support is of course better provided by a significant staff of knowledgeable scientists and development engineers working full time to code enhancements and support in the Firm marketing the code than by other solutions.

An estimation of the quality of the support team can usually be performed looking at the quality and completeness of the code documentation. A good level of support usually implies high quality manuals, that tend to be more complete, readable and useful.

Furthermore, because of the very high cost of development of large software codes and packages in general, they are usually marketed as soon as they are provided with the minimum amount of features that makes them usable. Further developments and addition of features are generally made with the code already on the market, also taking into account the feedback of information of practical users.

From the potential user standpoint, this makes significantly more appealing, main features being equal, a Firm provided with a significant development staff and well established on the market, that is likely to provide a more stable environment and has better chances to continue the marketing and support of the code for the whole depreciation period.

Finally, the location and staff of the nearest assistance station should be of course also evaluated.

4.5 COST STRUCTURE

It is not certainly to be stressed, since it appears absolutely obvious, particularly for industrial applications, that costs are an important evaluation parameter. The aim of this subsection, however, is to provide indications for a correct comparative evaluation of costs, since for software products their components are not always obvious or apparent, and they can be provided in a variety of forms.

To perform correctly the evaluation of costs, the potential buyer should check that the figure of cost, besides the cost of the analysis code, pre- and post-processor, include the following items, necessary or highly advisable:

Cost of ancillary software. This certainly includes the graphic library and the drivers of the graphic devices, that may or not be included in the basic price, and could also include rental fees of other proprietary software embedded in the code.

Cost of installation. Even if the code Vendor usually provide the alternative of installing the code directly or send a tape and have it installed by the user, it is generally highly advisable in an industrial environment, in the experience of the author, to have the code installed by the code Vendor. This is particularly true when the hardware is not a fully standard one and for large 3D codes. Installation by the code Vendor, even if generally requiring an extra cost, allows to determine timely potential problems and to minimize the time required for a full operativeness.

Cost of documentation. One or more full set of documentation are obviously required, according to the number of users envisaged for the code. The cost of documentation should be then clearly defined, since a variety of policies are applied by code Vendors in this respect.

Cost of specific hardware. If the hardware is to be bought and used exclusively to run the code, its cost should be added directly to that of the code. Of course a variety of different situations are possible, and the fraction of hardware cost to be charged to the code should be defined accordingly.

Guarantee and initial maintenance. This cost is included in most cases for an initial coverage of one year. However, since the policies of the various Vendors can vary significantly also in this respect, a precise knowledge is necessary to perform fair comparisons.

Particularly for large codes, it would be advisable to include also an instruction course, and have an indication of the cost of maintenance after the expiration of the guarantee.

Once the cost components are all known for all the codes under examination, a "normalization" of costs is necessary, since some costs will be rental fees and other total costs. Furthermore, also the main analysis code can be found priced both ways, or even with leasing strategies, and some Vendors do not provide a choice between different pricing policies.

In order to perform this "normalization", it is then necessary to define a depreciation rate of the code. As an indicative criterion, that should however be checked against possible different company requirements or policies, a depreciation rate of 20 to 30 percent per year is usually assumed for this type of investment.

5. Survey of Market Situation

In this Section it will be provided the information for a very synthetic overview of the codes and packages known to the author to date as marketed to cover some aspect of electromagnetic analysis.

Because of the very dynamic market situation previously outlined, the list is necessarily bound to be incomplete; even for the listed products, the sketchy indications of coverage and hardware supported are also likely to be somewhat out of date and probably incomplete.

This information should then be used only to provide orientation at large of some of the codes on which to focus attention, as a function of the user's requirements, to perform further detailed comparative analyses, according to the guidelines stated in the previous Sections.

Further searches in the technical literature are anyway always advisable, since they can provide a more precise feeling of the activity and background of each code developer, and also give indications of codes officially available on the market and not mentioned in this paper. In this case, a note to the author, directly or through the marketing Firm, would be highly appreciated.

The list is given in Appendix I, and for each code are given, when known and applicable, the geometries handled, the method used, the type of hardware supported, the classes of problems covered as well as the name of the main code Vendor. Codes are listed in alphabetical order of code Vendors, to group together all codes marketed by the same Firm, so that the order has no significance. The abbreviation used are decoded in the list of abbreviation also given at the end of Appendix I.

Finally, in Appendix II are listed the addresses of code Vendors mentioned in Appendix I, including the means known to the author to contact them. Several Vendors can have representatives or distributors, whose address could be obtained, when of interest, by the main Vendor directly.

When a code of interest has been identified, and the evaluation procedure has been completed, a further decision should be made between simple access and acquisition of the code.

To have simply access to the code, through the company marketing the software or through a consulting Firm, is usually possible for most of the codes present on the market, even if the location where the service can be obtained should be also considered, and it is not necessarily bound to be practical.

When possible in practice, this approach, in the experience of the author, could be of interest when an occasional use only of the code is necessary, and then full acquisition does not appear advisable or economically feasible.

A second case that could suggest such approach, this time on a more temporary base, is when a deeper technical evaluation than that possible with benchmarks is considered necessary to evaluate in more detail the matching between code capabilities and user requirements. In this case it is usually also advisable to have the technical personnel of the commissioning Firm following the analysis to be performed, in conjunction with the code Vendor or the consultant, in order to gain a deeper direct insight of the code capabilities.

However, when a code has proven suited to the user's needs and the envisaged use is potentially significant, direct acquisition is generally advisable, also taking into account the rapid increase of application areas covered and the constant need, in most industrial areas, to improve design features to maintain competitiveness.

6. Conclusions

In the opinion of the author, it can be stated as a conclusive remark that Computer Aided Electromagnetic Analysis has reached an "initial maturity" stage of development, and can offer a significant set of commercially available and reliable tools of interest for the design of electromagnetic devices in an industrial environment.

A significant number of "success stories" of industrial application have been already registered, and these kind of codes have gained acceptance as essential tools for the design of advanced devices such as high energy physics dipoles and quadrupoles, electrostatic lenses, magnets for medical imaging, experimental fusion reactors and magnetic transducers and actuators.

An increasing awareness of the importance of these analysis tools is also developing for more traditional industrial applications, in all cases in which the quality of design already is, or is becoming, important to improve performance and/or to reduce costs.

Because of the difficulties in a complete physical modelling of the materials involved and of the intrinsic complexity of the area, coverage of industrial applications is not complete yet, but is rapidly improving and appears well backed by research and development activity.

Since the codes are relatively complex to use, particularly in an industrial environment, because of the three-dimensional geometrical complexity of most electromagnetic devices of practical interest and of the intrinsic difficulties previously mentioned, a policy of attention to the area and of early acquaintance with the code usage is advisable in all cases in which a potential interest is envisaged.

7. Acknowledgements

Most of the views presented in this paper are the result of many years of research activity in the "Laboratorio di Progettazione Assistita da Calcolatore di Dispositivi Elettromagnetici" or briefly PACDE Lab, of the Electrical Engineering Department of the University of Genova.

The author is consequently deeply indebted with a large number of persons that have worked with him over these years, and particularly with Paola Girdinio, Paolo Molfino, Maurizio Repetto and Alessandro Viviani, with which it has been established a long and fruitful scientific relationship.

APPENDIX I

List of software for electromagnetic analysis
known to the author as officially marketed
in July 1990

Code name	Geom.	Method	Hardware	Coverage	Vendor
MAXWELL	2D,AS	FE	P,W...	ES,MSnl,ECsstr,HF	Ansoft Corp.
MAXWELL 3D	3D	FE	P,W...	ES,MSnl	"
FLUX2D	2D,AS	FE	W,m,M,N	ESK,MSnl,ECsstr, Th,CTh,...	Cedrat SA
FLUX3D	3D	FE	W,m,M	ES,MSnl	"
PREFLU	2D,AS	–	W,m,M	Prepr.	"
PHI3D	3D	BE	W,m,M	ESK,MS	"
Microflux	2D,AS	FE	P	ES,MSnl,ECss	"
CAD/CEM	–	dpc	W,m,M	EC,HF	"
MAGNA/FEM	2D,AS	FE		ES,MS...	CRC Corp.
MAGNA/FIM	3D	FE		ES,MS,.	"
MAGNA/ELF	3D,AS	CS		ES	"
MAGNA/JIBA	2D,AS,3D	Mo		MSnl...	"
FLUX-EXPERT	3D,AS,3D	FE	W,m,S...	Gen. Pur. Sol.	DT2i
MAGNUS-3D	3D	FE	W,m,M...	ES,MSnl	Ferrari Ass.
GE2D	2D,AS..	FE		ES,MSnl,ECsstr...	General El. USA
PDE/PROTRAN	2D,AS	FE	W,m,M...	GeN. Pur. Sol.	IMSL
MagNet	2D,AS	FE	P,W,m,S	ES,MSnl,ECsstr...	Infolytica
MagNet3D	3D	FE		ES,MSnl	"
ELECTRO	2D,AS	BE	P	ES	Int. Eng. Soft.
MAGNETO	2D,AS	BE	P	MS	"
COULOMB	3D	BE	P	ES,pt	"
JMAG	2D,AS,3D	FE	m,M...	MSnl,ECtr,V...	JAIS
Maxwell	2D...	FE	P	ES,MSnl	MAGSOFT Corp.
MSC/EMAS	2D,AS,3D	FE	W,m,M...	ES,MSnl,ECsstr,HF	MacNealSchwend.
MSC/XL	2D,AS,3D	–	W,m,M...	Prepr.	"
MSC/MAGGIE	2D,AS	FE	P	ES,MSnl,ECsstr,Th	"
PETFEM	2D,AS	FE	P	ES,MS...	Princeton E.T.

Code name	Geom.	Method	Hardware	Coverage	Vendor
PROFI	2D,AS	FD,FE	m,M...	ES,MSnl,ECss,Th...	Profi Eng.
	3D	FD,FE	m,M...	ES,MSnl...	"
COSMOS/M	2D,AS	FE	P,W	MSnl,ECsstr,Th	Str.Res.An.Crp.
TEAP	3D	FE		MSnl,Th	"
ANSYS	2D,AS	FE	P,W,m,M	ES,MSnl,ECSStr...	Swanson An.Sys.
	3D	FE		ES,MS...	"
MEGA	2D,AS,3D	FE	W,m,...	ES,MSnl,ECsstr,VDV,HF	Univ. of Bath
PE2D	2D,AS	FE	W,m,M,S	ES,MSnl,ECsstr,VDV,HS,pt	Vector Fields
OPERA	3D	–	W,m,M,S	Prepr.,Postpr.	"
TOSCA	3D	FE	W,m,M,S	ES,MSnl,pt	"
ELEKTRA	3D	FE	W,m,M,S	ECsstr	"
GFUN	3D	VIE	W,m,M,S	ES,MSnl	"
BIM2D	2D	BI	W,m,M	MS	"
TAS	2D	TS	P	ES,MS	"
WEMAP	2D,AS	FE	W,m...	ES,MSnl,ECsstr...	Westinghouse

LIST OF ABBREVIATIONS

used in Appendix I

Method
- FE = Finite Element
- BE = Boundary Element
- VIE= Volume Integral Equation
- Mo = Moments
- Prepr. = Preprocessor
- FD = Finite Difference
- BI = Boundary Integral
- CS = Charge Simulation
- TS = Tubes and Slices
- Postpr. = Postprocessor

Geometry
- 2D = Two dimensions
- AS = Axisymmetric
- 3d = Three dimensions without symmetries

Hardware
- P = Personal Computer
- m = mini/superminicomputer
- N = Accessible through networks
- W = Workstation
- M = Mainframe
- S = Supercomputer

Coverage
- ES = Electrostatic
- MS = Magnetostatic
- EC = Eddy Currents
- tr = general transient
- CTh= Coupled with thermal
- VD = Voltage Driven
- HS = Hysteresis
- ESK= Electrost. and Electrokinetic
- nl = magnetic nonlinearity
- ss = sinusoidal steady state
- Th = Thermal problems
- HF = High Frequency eddy currents
- V = Velocity Effects
- pt = particle trajectory
- Gen. Pur. Sol. = "General Purpose" equation Solver

General
- ...= possibly incomplete
- = unknown
- – = not applicable
- " = same as above

APPENDIX II

Full names and addresses as known to the author
of Electromagnetic Analysis code Vendors
mentioned in Table I

ANSOFT CORPORATION
Univ. Techn. Devel. Center
4516 Henry Street
PITTSBURGH, PA 15213   USA
Phone: 412-683-4846

CEDRAT
Chemin du Pre' Carre'
ZIRST
38240   MEYLAN   France
Phone: (76) 90.50.45

CRC - Century Research Center
Head Office
3-6-2 Nihonbashi-Honcho, Chuo-ku
TOKYO   103   JAPAN
Phone: (03) 665-9711

DT2i-Develop. and Transfer
in Inform. Ind.
Chemin des Preles
38240 MEYLAN ZIRST France
Phone: (76) 410510

FERRARI ASSOCIATES, INC.
P.O. BOX 1866
ORANGE PARK, FL 32067-1866   USA
Phone: 904-282-6041
Fax: 904-282-3910

GENERAL ELECTRIC COMPANY
Dr. M.V.K. Chari
Electromagnetics Program
Corporate Research and Dev.
1, River Road,
Schenectady, NY 12345   USA
Phone: 518-3875208

IMSL
2500, Permian Tower
2500 CityWest Boulevard
HOUSTON, TEXAS 77042-3020   U.S.A.
Phone: 713-782 6060
Fax:713-782 6069

INFOLYTICA Corp.
1140 De Maisonneuve,
Suite 1160
MONTREAL h3A 1M8 CANADA
Phone: 514-849-8752
Fax: 514-849-4239

INTEGRATED ENGINEERING SOFTWARE
347-435   Ellice Avenue
WINNIPEG, MANITOBA R3B 176
CANADA
Phone: (204) 942-5636
Fax: (204) 944 8010

JAIS-JAPAN INFORMATION SERVICE LTD.
CAE Dept.
3-5-12 Kita Aoyama, Minato-fu
TOKYO, 107   JAPAN
Phone: 03-475-0917
Fax: 03-423-8966

MAGSOFT Corporation
1223, Peoples Avenue
TROY, N. Y. 12180 U.S.A.
Phone: (1) 518-271-135

Mc Neal Schwendler Corp.
815 Colorado Boulevard
Los Angeles, CA 90041-1777 U.S.A.
Phone: 213-258 9111
Telex: 4720462
Fax: 213-2593838

Princeton Electro-Technology, Inc.
5815 Doverwood Drive, Unit 1
Culver City, California 90230
U.S.A.
Phone: 213-337-1716

PROFI ENGINEERING
Prof. Dr. W. Muller or
Dr. Ing. U. Hamm
Wilhelminenstr. 21
D-6100 DARMSTADT FRG
Phone: (06151) 264-18

Structural Research and
Analysis Corp.
1661 Lincoln Boulevard, Suite 100
Santa Monica, CA 90404 USA
Phone: 213-452-2158
Telex: 705578
Fax: 213-399-6421

Swanson Analysis
Systems, Inc.
Johnson Road, P.O. BOX 65
Houston, PA 15342 USA
Telex: 510-6908655
Fax: 412-7469494

University of Bath
Prof. J.F. Eastham
or Dr. D. Rodger
School of Electrical Engineering
BATH, AVON BA2 7AY UK
Phone: (0225) 826056
     or (0225) 826052

VECTOR FIELDS LTD.
24 Bankside
KIDLINGTON
OXFORD OX5 1JE U.K.
Phone: (08675) 70151
Telex: 83147 VIAOR G
Fax: (08675) 70277

Westinghouse Research Center
Mr. William S. Woodward
WEMAP Manager
1310, Brulah Road
PITTSBURGH, PA 15235 U.S.A.
Phone 412-256 1665

GUIDELINES FOR AN EFFECTIVE USAGE AND PROCEDURES OF VALIDATION

GIORGIO MOLINARI
Dipartimento di Ingegneria Elettrica
Universita' di Genova
11a, Via Opera Pia
I-16145 Genova Italy

ABSTRACT. In the paper a series of guidelines, based on the author's experience, for an effective exploitation of Computer Aided Electromagnetic Analysis tools in an industrial environment will be presented, and the most significant procedures and initiatives for the validation of this class of codes will be discussed.

1. Introduction

Once an Electromagnetic Computer Aided Analysis code or package has been selected for use in the design department of an industrial Firm, and the installation procedure has been successfully completed, it is necessary to plan the integration of the new tool in the design procedures that could profit of its presence.

In the experience of the author, particularly when the use of the code should be extended to persons different from those that have evaluated and selected it, this integration can become critical, and requires a careful planning to be concluded successfully.

Even when the use of the code will be restricted to the persons involved in its selection, the full and correct exploitation of code capabilities is not generally trivial and should be carefully planned.

The importance of this phase is frequently underestimated, maybe also because of the "ease of use" boasted by some vendors, but the experience gained so far indicates that it is frequently crucial for a full and successful integration of the new tools in the design procedures.

In general, a careful planning of the code or package exploitation should consider a series of successive steps, each of which will be discussed in some detail in the following Sections.

Also extremely important is the control of the presence and efficiency of a sufficient provision of graphic devices and computational power, whose absence or deficiencies could lead to a poor and frustrating usage of the code.

However, this monitoring of the "hardware backing" of this kind of code or packages will be assumed as provided initially and continued in the following, and will not then be mentioned any more in the rest of paper.

Y. R. Crutzen et al. (eds.), Industrial Application of Electromagnetic Computer Codes, 101–120.
© *1990 ECSC, EEC, EAEC, Brussels and Luxembourg.*

2. The field approach

The classic design procedures of a large majority of electromagnetic devices, as already mentioned in the previous paper, is based mainly on a circuit approach, while procedures based on a field approach are generally rare and of limited use.

This is particularly true in an industrial environment, that usually takes advantage significantly of the greater simplicity and ease of use of circuit modelling, making a very wide use of "equivalent circuits" of electromagnetic devices.

When starting from a background with these features, an effective use of Electromagnetic Analysis tools can become difficult, and can require some brushing up of practical field theory and a careful analysis and appreciation of the design problems that could be better modelled in term of fields. An underestimation of this problem can lead to a poor utilization of the code or even to a complete inability to integrate it in the design cycle.

Of course, the importance of the problem can however vary very much in any specific case as a function of the background of designers, the type of products to design, human factors or other reasons.

The best approach to face the problem in large design Centres, in the opinion of the author, is to provide some internal short courses, to distribute liberally documentation on the subject and to make available expert persons providing on request advice and suggestions and collecting the feedback of field practice.

Once the design procedures that could benefit of the usage of the new design tool have been clearly identified, it is also important to maintain the pressure for an integration of the analysis code in the design procedures.

On the other hand, the importance of a deep knowledge of field theory should not be overemphasized; what is really important is to develop some physical understanding of fields as a detailed design tool, and an excessive stress on field theory, that is a rather involved subject, could even have the opposite result with respect to the one desired.

3. Mastering the code

Once the usefulness of a field approach for a better modelling of some aspects of design of electromagnetic devices has been accepted, a frequent side effect is to come to expect too much from the code available.

As already pointed out in the previous paper, commercial codes at present on the market cannot handle all problems of practical interest, and present a series of limitations both specific of the code and relevant to the general coverage of the area developed so far by the scientific community.

However, even with these limitations, many codes can already face successfully several classes of problems of interest both for research and industrial applications, and they are also rapidly extending coverage and user friendliness, as also detailed in the previous paper.

Capabilities and limitations of each code should then be clearly and timely explained to potential users at the initial phase of the introduction of the code, to orient correctly the types of applications, preventing possibilities of future disappointments, and to collect feedback of areas deserving coverage in design activities for future extensions.

In addition to this appreciation of capabilities and limits of the code, it is obviously also necessary some specific training with the procedures and specific commands of the code.

Any code of this type comes generally provided with a user manual, or even a set of manuals, and they should be studied carefully by potential users. However, the general experience with these, like many other, manuals is that they can be useful or very useful for the expert user, but are of limited help to the inexperienced one.

Consequently, it is advisable in most cases to plan a specific training on the usage of the code, particularly for the large 3D ones, usually provided with a richer set of commands.

This training should be a practical, "hands on" one, giving to the user the possibility to encounter and solve also a number of small practical difficulties usually associated with the management of any kind of large code or package of codes.

Even with this head start, a decreasing need for assistance is likely to be useful for some time; in large design centres it is then advisable to appoint a "code manager", trained at a higher level, able to provide most of the required support and responsible for the code maintenance and for the collection of feedback from the users.

For small users, this function could be performed through a "hot line" that should be established with the assistance service of the code vendor or with an user provided with larger experience.

4. Problem definition

The amount of computer resources required by the solution of an Electromagnetic Analysis problem is generally rather large, and is likely to be very large for three dimensional geometries. As a consequence, it is also likely that the analysis of three-dimensional geometries of practical interest, usually rather involved, become significantly compute-bound, and then require a careful definition of the problem to fit in the computer resources available.

Even when treating simpler problems, that can be easily hosted by the hardware available, it is anyway always important to simplify the problem to its minimum dimension of interest, to save computer resources, to reduce costs and above all to minimize the response time, that could play a significant role in a profitable integration in design procedures.

Once selected the aims of the computation and the accuracy required, the problem should be then defined with great care, exploiting all means available to minimize the computational burden.

A key practice in this respect is that of a proper definition of the geometry to handle for the computation, obtained avoiding unnecessary details while preserving the essential features required.

Of course, these abilities are obtained essentially with practice and experience, and should then be developed over a significant time, solving a number of problems and gaining confidence in the results obtained.

However, a few general guidelines to aid this process can be defined, and are detailed in the following:

Exploitation of symmetries. This is the first and maybe most important simplification that can be performed in a computation. Very often practical devices, particularly in 2D geometries, present some kind of symmetry, that allows to define lines or surface on which boundary conditions can be derived, and that can then be exploited to reduce the domain to be solved numerically to one half, one quarter or even one eighth of the original one.

A further reduction of the solution domain is frequently allowed by the possibility to impose "periodicity conditions", that is to define lines or surfaces were the solution, even if not known, is bound to be symmetric or antisymmetric with respect to a similar line or surface. This condition, however, requires a specific extension to be handled by the code.

The exploitation of symmetries, while reducing significantly computer resources and response times, does not affect the accuracy of solutions, and is then to be absolutely exploited in all possible cases as the first problem definition step.

Simplification of geometry. This is frequently a very important point to obtain faster and more efficient solutions. Practical devices of industrial interest are often of rather complex shapes, since they have to satisfy a number of constraints coming from a variety of engineering and technological fields. However, not all details of the device are equally important for each of the specific physical phenomena it should handle, and electromagnetics is no exception.

Consequently, areas not interested by electromagnetic phenomena could be left out or drastically simplified without affecting the quality of the electromagnetic analysis. More difficult to define, but often equally or more important, are simplifications to be performed in area interested by electromagnetic phenomena, but to such a limited extent as to be likely not to influence significantly the results of interest. In this case the evaluation of the possible simplifications to perform is of course more difficult, and requires experience, to be gained performing solutions, in doubtful cases, with and without the simplification under consideration and comparing results.

However, a simple "rule of thumb" that can profitably be used to provide indications in this respect is that simplifications are likely to be possible in areas far from, or weakly linked with, the main regions of interest.

Minimization of computational difficulties. A further consideration that should be also kept in mind by the user is that not all the geometries that can be defined, and not all features provided by the code, require the same computational effort.

On the geometric side, the use of sharp convex edges, that mathematically require an infinite field on the tip, usually generate difficulties in numerical convergence, and should then be kept to a minimum or avoided, replacing them with some kind of rounding when not essential for the computation.

As far as advanced features are concerned, such as complex constitutive relations of material, it should be considered that these features generally implies more complicate solutions and usually also less well conditioned situations, so that they are very likely to require a significantly larger computational effort. This of course is not to say that these features should not be used, but that their use should be made only where really required, avoiding to model in this way unnecessary details or parts not playing a significant role in the solution.

Modelling of open boundaries. When it is necessary to model open boundaries problems and the code does not provide an automatic coverage of this feature, it is then up to the user to define a finite line or surface modelling a "technical infinity", where it is felt that the main electromagnetic phenomena have negligible effects, and a boundary condition can then be placed. If this definition is made with too conservative assumptions, this could lead to unnecessarily large and time consuming problems to solve, while an unrealistically closed boundary defines and solves a different problem.

To make a correct selection, experience is of course the best solution, and the user should gain confidence in this respect solving an initial series of problems of interest of this type with several different boundaries, to select the best compromise for its kind of devices.

As a general initial guidance, its should be considered that a zero boundary condition can generally be introduced safely at a distance of the order of a few times the maximum dimension of the device of interest.

5. Meshing

The most diffused procedures for numerical computation of electromagnetic fields require some kind of discretization in the domain of interest. Even if several of the remarks defined in this Section apply to a variety of methods, at least partly, reference is made particularly to the ways of facing meshing problems for the Finite Element Method, at present the most diffused in Electromagnetic Analysis and, in the author's experience, in Computer Aided Analysis at large.

Finite Element, like most numerical methods, defines a discrete approximation of a continuous problem, then necessarily introducing discretization errors with respect to the exact solution. These errors are a function of the method used, of the type of problem, of the local structure of discretization and can also interact with other sources of approximation.

It can be shown, at least in well-behaved cases, that this approximation error decreases with the size of discretization, and converges to zero for an hypothetical infinitely fine mesh. This means that the level of approximation is linked with the discretization performed, and that it improves as the mesh is refined.

Consequently, as a general criterion, meshes should be finer where a higher accuracy of the solution is required, while they could be coarser when a lower level approximation is sought. This applies not only to the solution of different problems, but also to different subdomains of a single one, when the accuracy required can be clearly defined in each of the subdomains and the code allows the definition of graded meshes.

In turn, this implies that the user will influence the amount and distribution of discretization errors through the definition of any specific meshing of a problem.

In principle, particularly with the Finite Element method, discretizations are relatively "robust", so that the selection of a given meshing is not a very critical procedure, and many different meshes can provide acceptable results; however, the user should have a sound feeling of the "quality" of a given meshing, to be able to define or evaluate the correctness of a given mesh for the solution of a given problem, avoiding both inadequate approximations and excessive refinements ("overkills"), that reflects heavily on computer resources and time to solution.

With traditional codes, this requires a significant experience, even when semiautomatic or automatic meshing facilities are provided, since the ultimate evaluation of the quality of the mesh is always left to the judgment of the user, on the basis of the accuracy required, of which however no quantitative estimation is given by the code.

In the following will then be given general guidelines to help the user to define acceptable meshes and to build up a confidence in the "quality" of his/her meshing. More advanced solutions now starting to be available, based on error estimation procedures, will be dealt with at the end of the Section.

To control the quality of a mesh in order to obtain an acceptable solution, a "top-down" approach starting from the desired design results should be defined. The main steps to follow, according to the author's experience, are detailed below.

Definition of outcomes required. This is a very important preliminary step to be performed, since the type of outcomes desired influence significantly the definition of the meshing required.

Accuracy being equal, a solution out of which only integral parameters are required can have a significantly rougher meshing, with the possible exception of computation of forces and torques. In fact, even a rather crude approximation of electric and magnetic fields will tend to balance out, providing improved accuracy, if only quantities such as inductances, losses, capacitance, requiring an integration of fields are to be computed.

On the contrary, when maximum or minimum values of field quantities are to be evaluated, a much more refined meshing is required to obtain the same level of approximation. This need of mesh refinement is even stronger when, as often happens in Electromagnetic Analysis, the solution of the problem is obtained by means of a potential. In fact, in this case the electric and magnetic fields must be derived from the solution by means of numerical differentiation procedures, that are intrinsically bound to worsen the accuracy of results, particularly on boundaries and interfaces and where steep variations occur.

The user should then define before starting to mesh which kind of outcomes are required, to have indications on the type of meshing necessary.

Definition of the accuracy desired. Once the above general indications on the type of outcomes have been defined, a second key decision to be made for a proper selection of the mesh is relevant to the accuracy desired.

Of course, an high accuracy requires in general finer discretizations and then larger problem and longer solution times, so that the accuracy required should not be kept too high when not necessary.

A problem arising in this context however is that, when the required accuracy has been selected, the number of unknowns, or degrees of freedom, necessary to obtain this accuracy in a given computation should be evaluated; in traditional codes not provided with error estimation procedure, like most of those presently in the market, this decision is to be made essentially on the basis of the user's experience.

In fact, it is not easy to provide significant suggestions on this topic "a priori", since the optimum number of unknowns (and then of nodes), even when the type of outcomes has been defined, still depends on many variables, such as the type of geometry and the shape of the solution domain, the number and type of materials to model, the method of solution and the type of elements, and the steepness of field variations.

However, to provide an initial broad guidance to the user in the selection of the order of magnitude at least of this figure, it could be stated that for an "average" 2D problem are usually required from some hundred to a couple of thousand nodes (and then unknowns in 2D), mostly according to the type of outcomes desired and to geometrical complexity, to obtain an "average" accuracy, with errors of a few percent.

However, these figures are bound to increase rapidly in 3D: if we assume as 200 and 2000 the lower and upper bound of the range of nodes necessary to model an "average" 2D problem on a "square" domain, and we suppose to be willing to model a "cubic" domain with the same average mesh size, we could have a rough estimate of the number of nodes required by multiplying the previous numbers by their square root, thus getting figures of the order of 3,000 and 90,000, respectively.

Furthermore, this applies only to magnetostatic and electrostatic problems, that can be modelled with a single unknowns per node; for magnetodynamic or eddy current problems some regions at least will have to be modelled with more than one, and up to four, unknowns per node, thus increasing significantly the number of unknowns required.

To try to cope with this situation, usually, for initial preliminary design at least, the accuracy required in 3D is then somewhat relaxed, allowing then some reduction in the node number.

Since the most appropriate node number is anyway strongly dependent on the application, to obtain a deeper appreciation of the approximate node number to select, each user should perform a series of tests on some "standard" problems of its class with different node numbers, comparing then accurately the results. In this way it can be obtained a more direct and reliable indication of the range of node number to select for any specific class of applications.

Definition of local consistency of meshing. When the general structure of the mesh has been defined, it is necessary to select local meshing criteria, or perform a series of checks of local mesh consistency if an automatic meshing procedure has been used. This is particularly important when an evaluation of minima and maxima of field quantities are required, since in this case a significant refinement with respect to the average mesh size in the areas where the maxima are likely to be located is usually advisable.

A few general criteria for this kind of operations can be given:

- Mesh should be fine enough to model properly all geometric details of interest. With first order elements, it is advisable to model thin subregions with at least two elements covering the minimum thickness.

- In each area, mesh should be refined where the fields of interest are steeply varying. On the contrary, mesh can be significantly rougher than average where the fields are nearly constant.

- With triangular finite elements, mesh size being equal, discretization errors are minimized if triangles are nearly equilateral.

An example of the difference in meshing produced by different strategies of use of the same overall node number is given and commented in Fig. 1 in the next page, for an electrostatic case in which the computation of the electric field between two slightly different needle electrodes was required.

With the above set of guidelines, users should generally be able to build up rather quickly an acceptable confidence in their ability to mesh adequately problems of industrial interest. As experience of use develops, guidelines will be less and less necessary, and the user will learn to define and build rapidly adequate meshes on the basis of experience alone.

Even if meshing ability is at present one of the key features to make an optimal use of Electromagnetic Analysis codes and maybe one of the most significant difficulties that hinder a larger diffusion in industrial applications, this is likely to become less and less the case in the future.

In fact, a few codes are now starting to offer, mostly in 2D so far, error estimation and adaptive meshing algorithms, that can provide in principle a solution within an accuracy specified initially by the user.

This would allow, at least in principle, to the user to specify the required accuracy and then ignore completely the meshing procedure, being anyway confident that the adaptive meshing will provide a solution with the accuracy desired.

Of course, this also means that the problem will have to be solved automatically several times, refining the mesh where indicated by the error criterion until the error estimate required is satisfied everywhere, so that the computational burden is bound to be significantly larger. However, the recent release of advanced workstations, based on vector and parallel architectures and able to provide a significant computational power at acceptable prices also for CAD applications, makes this approach promising of developments of very significant interest for industrial applications in the near future.

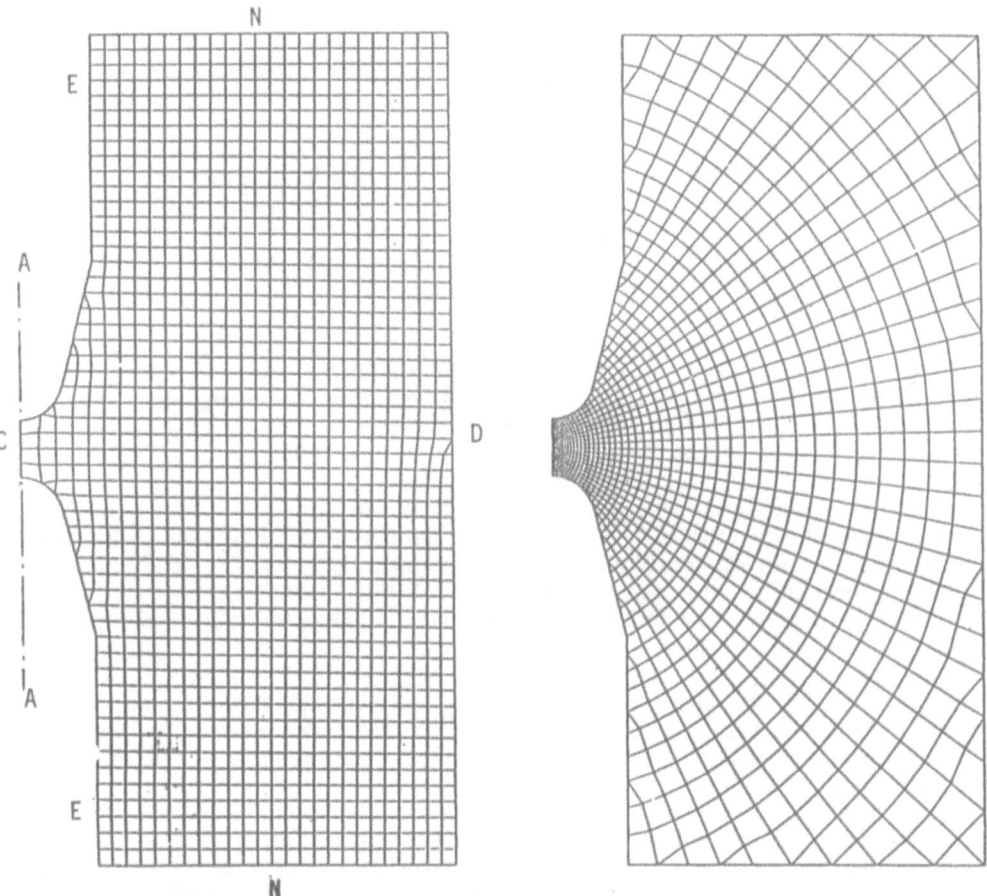

Fig. 1 – Example of different meshes with the same number of nodes but different distributions for the meshing of an electrostatic computation relevant to the evaluation of the electric field between two needle shaped electrodes. 1500 nodes in both cases.

A – A Axis of symmetry
E – E Needle electrodes
N – N Boundaries with Neumann condition
D Boundary with zero Dirichlet condition

In this case, the importance of the computation is concentrated in the evaluation of the electric field in the gap between the electrodes.
The mesh on the left, built up with an uniform distribution of nodes, will provide a very limited number of nodes in the important gap region, where the electric field is steeply varying, and then a poor accuracy.
Much better suited to the problem is the mesh on the right, that presents a well refined mesh in the gap, and a coarser one in the less important external region.

6. Control of the solution

In spite of any training and even when a significant experience has been gained, users of large codes, such as the Electromagnetic Analysis ones, are very likely to make mistakes.

In fact, the complete definition of a problem requires a significant number of inputs with a variety of commands, so that the probability of errors in defining for instance geometric coordinates or material parameters is generally significant. Further possibilities of errors arise in the selection of options, that can be correct in principle, and then pass undetected through error checking routines, but inappropriate in the context of a specific problem.

Errors readily detected and pointed out by the error checking and recovery procedures of the code, even if disappointing and time consuming, are the less dangerous ones, since do not lead to wrong outcomes.

Also not very dangerous are errors leading to impossible or obviously wrong results, readily detected and corrected by the user.

Much more dangerous are errors arising from an improper but in principle correct use of the code and leading to wrong but possible results; also very dangerous are errors producing a well defined and correct problem, but different from the one the user expects.

Further errors can then arise from inadequate meshing, or from code bugs or inadequate software/hardware environment.

Because of all the above possibilities, the user should then assume, in the opinion of the author, a rather "suspicious" attitude with respect to any new computation, and try to verify the consistency and quality of the solution with all means available.

Some guidelines to check the consistency of results for two-variable problems (that is, two-dimensional or axisymmetric) are given in the following subsections, for magnetic and electric problems, respectively; a pair of relatively "standard" types of errors that can be detected with a careful examination are also presented and discussed in Figures 2 and 3.

6.1 CHECK OF MAGNETIC SOLUTIONS

In magnetic two-variable problems solved for the vector potential **A**, only the z component turns out to be nonzero, so that the potential reduces to a scalar quantity, and equipotential lines are flux lines of magnetic induction **B**. In these cases, plots of lines of constant vector potential are usually provided by any code, and the following properties hold:

- Equipotentials cannot cross each other nor a boundary where a constant potential is assigned (Dirichlet condition).

- Equipotentials should be perpendicular to boundaries where a zero normal derivative of the potential is specified (Neumann condition).

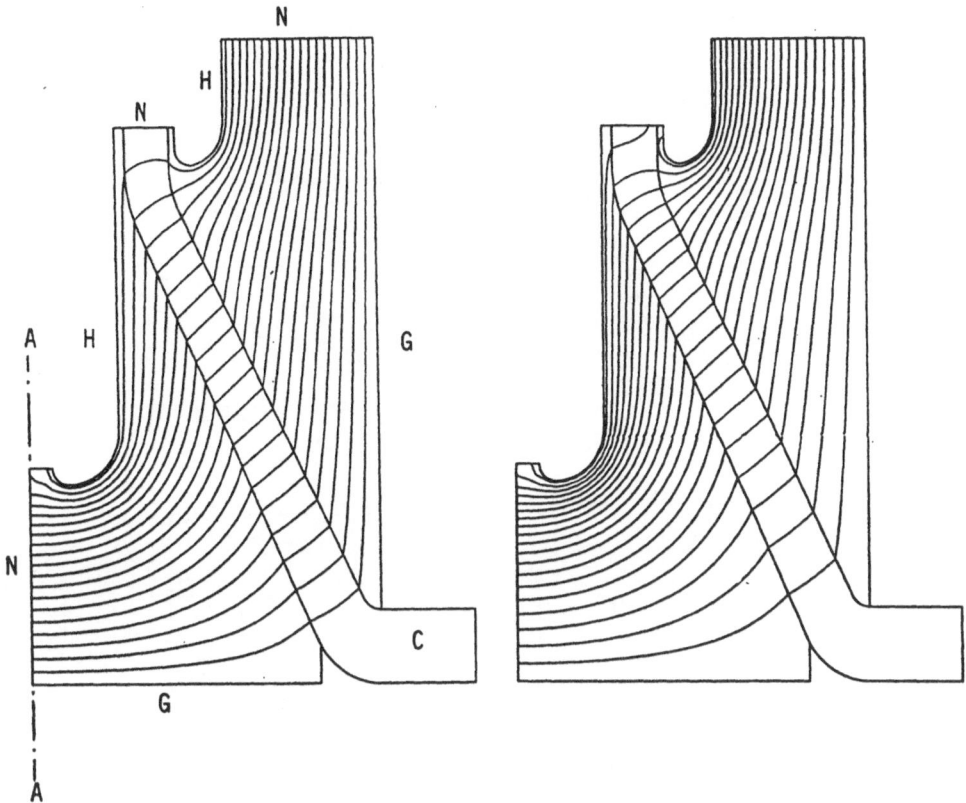

Fig. 2 - Identification of a data input error in the computation of equipotentials for an SF6 switchgear.

A – A Axis of symmetry
H – H High voltage electrodes
C   Insulating resin and fiberglass cone
G – G Grounded electrodes
N – N Boundaries with Neumann condition

The computation containing the data error is that on the left, whereas the correct one is given on the right for comparison. The error consists in the fact that, even if the real device is an axisymmetric one, the user, once defined the geometry, solved the problem with the option appropriate for plane case, thus selecting the wrong differential equation.
Even if the two computations appear very similar to each other, a close examination cna allow to identify the mistake. In fact, an axisymmetric solution is bound to have, in the upper region, practically cylindrical, an higher field on the inner electrode, with smaller curvature, and then equipotential more closely spaced than on the external boundary. The uniform spacing of equipotential in the solution on the left indicates a constant field, correct for a plane geometry, but not for axisymmetric cases.

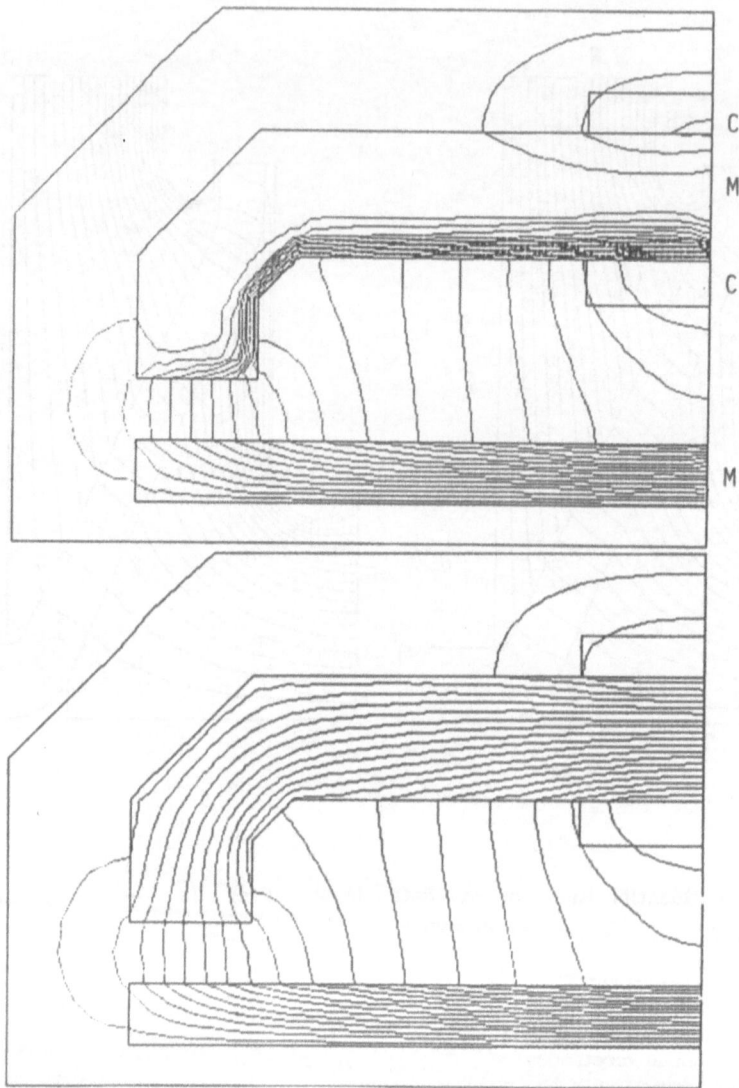

Fig. 3 - Identification of a code error due to inadequate convergence in a
2D magnetostatic computation of flux lines in a magnetic actuator.

C – C Exciting coil   M – M Magnetic material

The wrong solution, on top, can be identified since it does not present an
uniform distribution of flux lines at the coil center, where a uniform field
is expected.
The error in this case was due to the selection by the user of simple
precision for the main coefficient matrix, that for the relatively high number
of nodes was not providing enough significant digits for a proper convergence
of the solution algorithm used.

- The density of equipotential lines, when traced with the same spacing, is proportional to the local intensity of **B** (Ratio of field values can then be estimated).

- At an interface between two media with different magnetic permeability and no surface current, equipotentials, being also flux lines of **B**, are bound to be discontinuos. The ratio of the tangents of the angles between flux lines and the normal to the surface on the two sides of the interface should be equal to the ratio of permeabilities.

- Flux lines of **B**, when totally contained in the solution domain, must be closed lines linking some current or a part of a permanent magnet.

- When these closed lines are linked to a known current, a quantitative local check of the solution, once the magnetic field **H** has been evaluated, can be performed approximating the relation:

$$\int_l \mathbf{H} \cdot \mathbf{dl} = \int_S \mathbf{J} \cdot \mathbf{dS}$$

where S is the surface defined by the close line l.
This control is meaningful provided that the method used does not impose locally this condition.

- A flux line of **B** should show a discontinuity in the derivative when entering a region with nonzero current density. This condition holds irrespective of the nature of the materials involved.

6.2 CHECK OF ELECTROSTATIC SOLUTIONS

In electrostatic two-variable cases solved for the scalar electric potential V, the electric field **E** is perpendicular to equipotential lines at any point in the solution domain. In these cases, plots of equipotential lines are generally the first outcome, and the following properties can be derived:

- Equipotentials cannot cross each other nor a boundary at assigned potential (Dirichlet condition).

- Equipotentials should be perpendicular to boundaries where a zero normal derivative of the potential is specified (Neumann condition).

- The density of equipotential lines, when traced with the same spacing, is proportional to the local intensity of **E** (Ratio of field values can then be estimated).

- At an interface between two media with different permittivity and no surface charge, equipotentials are bound to be discontinuos. The ratio of the tangents of the angles between the normal to the equipotential in the two media and the normal to the interface should be equal to the ratio of permittivities.

- Equipotential lines must start and finish on boundaries with Neumann conditions or encircle completely a region with a Dirichlet condition.

- In axisymmetric problems, electrodes with smaller radius of curvature must produce on the surface an electric field of greater magnitude than electrodes with a larger radius.

- A quantitative check of the solution, when the electric field **E** has been computed, can be performed approximating the relation:

$$\int_l \mathbf{E} \cdot \mathbf{dl} = V_A - V_B$$

where the line l cover the distance between two boundaries, A and B, with different Dirichlet conditions.

6.3 GENERAL CONTROL SUGGESTIONS

Even if the properties given above could appear difficult or intricate to understand when explained verbally, they usually turn out to be self evident and informative in graphical form, once a little experience on the behaviour of fields has been gained.

This is the reason why, even if equipotential plots do not provide usually information directly useful for design, it is generally advisable to have them produced by default to accompany any 2D or axisymmetric solution.

An experienced user can extract a great deal of information at a glance from such a plot, and the beginner should constantly examine them to get acquainted with their features.

In three dimensions, the magnetic vector potential has all three components nonzero in general, and equipotentials of magnetic vector potential **A**, of magnetic scalar potentials, when defined, and of electric scalar potential V are always three dimensional surfaces rather then lines.

These makes of course graphical representation much more difficult both to produce and to understand, and breaks down any obvious link between equipotential and flux lines in magnetic cases.

The rules defined in the previous sections generally holds, but since graphical three dimensional representations of potentials are usually not available or not meaningful, checks of the solution with the rules of the previous sections are generally more difficult.

However, the designer of a class of devices always has some understanding of its expected physical behaviour, such as approximate location and expected range of maximum value of fields, distribution of eddy currents in the cross section of a damper, and so on.

He/she should then use this knowledge to verify that the solution provided by the code lies within the reasonable range of parameters that can be estimated, and shows an acceptable behaviour, matching the expectations, of all variables for which some pattern can be defined in advance.

This is frequently the case, for instance, of integral parameters, such as inductances, capacitances or losses, and of the behaviour of absolute values of fields in sections or along lines of the device.

All these quantities can generally be evaluated in many codes by means of the post-processor, and can then provide very useful information for a check of three dimensional solutions.

From plots of absolute values of fields and potentials, useful, though qualitative, information can also be derived in general on the accuracy of the solution: to help the user to develop this sensitivity, a few general remarks valid for any kind of equi-value curves can be stated:

- Since potentials and fields are bound to be continuous except at interfaces, the more one such line is jagged in a continuous region, the poorer the approximation.
  The reverse is also true, provided that no smoothing functions on the original solution have been used.

- Jagged lines or even small inconsistencies are not meaningful nor dangerous in areas where the variable being plotted assumes values close to zero. In fact, in these cases what is plotted is just the numerical "noise" present in any numerical computation when the variables approaches zero. A similar behaviour is to be expected anywhere if the solution is "zoomed" beyond the number of meaningful digits provided.

7. Validation

The procedures described in the previous Section are very important to allow to the user control on the quality and consistency of the outcome of an Electromagnetic Analysis code, but they do not provide in principle quantitative information on the accuracy of a given numerical solution with respect to the theoretical one of the differential equation, and even less with respect to the physical reality of the device modelled.

However, when a computation is really important for the evaluation of some critical parameter, the designer needs some sounder quantitative confirmation than the consistency checks described in the previous Section.

These procedures, generally called "validations" of the results computed by a code, are essential to build a real confidence in the code performance, and thus to complete a fruitful integration in the design process.

Several types of validation can be defined and performed; a classification of the main types, and a discussion of the features of each of them, is given in the following subsections, whereas in the final ones general validation initiatives and possible strategies for industrial validation of codes are briefly discussed.

7.1 COMPARATIVE VALIDATIONS

By "comparative validation" it is meant a procedure that allows to obtain further data on a problem already solved with a given code by means of other numerical solutions.

This is of course the least decisive type of validation, but is certainly the easiest to perform in most cases, and sometimes the only one possible.

This type of procedure can be further subdivided in a few subclasses, as detailed in the following.

7.1.1. Validations with the same code.
The simplest example in this class is a new run of the same problem with more nodes; as already mentioned before, this allows a control of the discretization errors of the original run.

A further type of validation in this class is to perform a different modelling of the device, allowing a sort of "sensitivity analysis" of the modelling assumptions testable with the code.

Anyway, this is the weakest possible type of validation, and could even be considered outside its proper definition, since it does not allow to test the basic modelling assumptions of the code nor its implementation.

7.1.2. Validation with a similar code.
This kind of tests, that can be performed in both ways described above, allows a validation of overall code correctness, method implementation and discretization of the previous run, of course provided that both codes produce reasonably close results.

This procedure also allow a control of data correctness of the first run, since the usually different procedures for data input make it unlikely to produce twice the same error in different forms.

However, since the two code have a similar coverage, this procedure cannot test the basic modelling assumptions common to both codes.

7.1.3. Validation with a more powerful code.
A standard procedure of this type is to test the results obtained with a 2D code with a second one having 3D coverage, and then able to account for the finite length or the real device.

This kind of procedure can provide the same kind of validation of the previous one, but also allows to check the modelling assumptions that is possible to relax with the more powerful tool, and provides then a significantly better coverage.

In general, comparative validation is a mean to add some reliability to the results obtained, "filtering" either a series of possible inconsistencies typical of numerical simulation algorithms or user errors, and than can provide a control of "invariance" of a solution with respect to a specific code, to a method of solution, or to some modelling assumptions.

This procedure become of significant importance for design of critical and innovative devices, when no previous experience exists and prototyping is either too expensive or too time consuming, or both.

In these cases it is advisable, when reliability of results is really of importance, to use this kind of validation, if possible with very different codes, at least until one or more of the codes available have been validated with the more reliable procedures described in the following sections.

7.2 THEORETICAL VALIDATIONS

With this expression it is meant a validation of numerical results obtained with a code by comparison with the analytical solution of the same problem.

This procedure allows to validate the results of *any* specific numerical method, providing a guarantee that none of the several approximations implied in a numerical solution has a significant impact on the resulting overall accuracy, with respect to the analytical solution of the same problem.

This naturally implies that an analytical solution has to be available; unfortunately, this is seldom the case for problems of interest in industrial applications, that generally presents shapes too difficult or impossible to handle in analytical terms with the partial differential equations of Electromagnetic Analysis.

As a consequence, this type of validation is usually done in relatively simple problems, of which an analytical solution is available or can be worked out.

Even if obviously of less direct impact than validations on final solutions, these procedures are anyway of significant interest, when a meaningful analytical solution is available, and can be useful both to code developers and users.

In fact, they allow to developers to "calibrate" a number of adjustable parameters, such as convergence thresholds, type of precision required for different variables, and so on, while users can profit of them to become confident in the code performance and to gain experience of the kind of meshing required to obtain a given accuracy with the code.

7.3 EXPERIMENTAL VALIDATIONS

The term "experimental validation" is used here to mean validations of results obtained modelling the behaviour of a specific device with an Electromagnetic Analysis code by means of experimental measurements on the real device.

This is the most complete and reassuring type of validation, since it allows to test the correctness of the basic modelling assumption of the code not only with respect to the analytic solution of the problem modelled, *but with respect to physical reality.*

In fact, in any real device, as previously observed, various physical aspects, pertaining to several analysis disciplines, generally coexist and interact; even within electromagnetic phenomena alone, a number of simplifying modelling assumptions are usually performed.

In a comparison with experimental results, all these assumption are tested as a whole, together with the numerical approximations. A good matching of code and validation results, besides providing information on code correctness, also provide the additional reassurance that the modelling simplifications performed with respect to physical reality do not play a significant role for the specific application.

Of course, this implies the availability of experimental measurements, that in turn mean that the device modelled with the code should exist already, so that this type of validation will not be possible in all cases, as previously mentioned.

Furthermore, experimental measurements should guarantee an accuracy and reliability greater, or at least equal, than that required to the code, otherwise possible discrepancies would be difficult to trace back to code performance.

In turn, this means that this type of validation is generally expensive, and can be usually done in an industrial environment only when measurements have to be done or are available for other reasons, and can be used for validation purposes as well.

Before using experimental results for validation purposes, it is important to make sure not only that they are accurate and reliable enough, but also that *all* information of interest required as input by the code, such as precise dimensional values, B-H curve of the real magnetic material used and so on, are *available and just as accurate,* otherwise again it will be difficult to draw significant conclusions on code performance.

7.4 GENERAL VALIDATION INITIATIVES

The problem of validation of Electromagnetic Analysis codes and packages is not of course a new one for the scientific community working in the area, and specific validation initiatives have taken place and have been reported in the literature over the years, generally associated with the presentation of a new or novel solution method or a new code.

The need of such kind of procedures, that could assure the user on reliability and accuracy of codes, is of course much larger, as pointed out before, for the development of large, innovative and critical devices, that cannot count on previous experience nor on prototyping to face challenging design problems.

Probably because of these reasons, the idea of starting an initiative for the definition of a series of general validation problems, independent from a specific environment, and allowing comparison of results with different codes, developed initially in the nuclear fusion energy community in 1985, particularly foe eddy current problems. After a positive response from the research community on Electromagnetic Analysis at the COMPUMAG Colorado Conference, a series of Workshops with biennial rounds, with Dr. Larry Turner of Argonne National Laboratory in charge as General Chairman, was launched.

The series of Workshops proposed a number of standard well defined problems, provided with analytical solution or experimental results, that were solved and compared by a number of researchers and code developers. The results obtained with the initial set of six problems of the first round was summarized in a special issue of the magazine COMPEL in 1988 [1]. The initiative, named TEAM (Testing Electromagnetic Analysis Methods) Workshops, was well received, so that a second round 1987-1989 was scheduled, always with Larry Turner in charge, and a summary of obtained results with six further additional problems was prepared [2], and individual problem was also made available on a special issue of the previously mentioned magazine [3].

Because of the continuing interest, the International Steering Committee of the series of Conferences COMPUMAG officially accepted to endorse the initiative, and launched a third series of Workshops, to be completed for the next COMPUMAG Conference, scheduled in Sorrento, Italy, for July 1991.

On the basis of the results of a questionnaire distributed among the scientific community, the scope of the series of Workshops was enlarged to cover all aspects of Electromagnetic Analysis, and some more additional problems were launched.

An extension of the advertising of the initiative is also planned, in cooperation with the Institute for System Engineering of the Joint Research Centre of the EEC at Ispra, that will also contribute to the preparation of a TEAM Newsletter. This is meant to increase the awareness of potential users and contributors of benchmark problems, particularly in the industrial environment, of the ongoing initiative, and to encourage them both to make use of the results and to participate to further developments.

Any enquiry or request of information on TEAM validation activities would be most welcome by the author of this paper, that is in charge of the initiative as General Chairman for the 1989-1991 round.

7.5 INDUSTRIAL VALIDATION STRATEGIES

Once an overview of the possible validation procedures has been set up, it is possible to discuss briefly which kind of strategy can be used by an industrial user to validate the codes of interest.

If the codes are to be used for the design of relatively "standard" products, of which previous experience and measurements are available, then, provided that the measurement possess the requisite previously described, it would be highly advisable to work out a series of internal validation runs on existing designs, in order to gain confidence in the code performance for the specific products of interest.

Even if time and resource consuming, this procedure would also allow to optimize the code use for the required performance, and then produce saving of resource very likely to pay back soon the investment, besides allowing a much more confident use of the code.

For the design of innovative devices, product-specific, internal validation procedures are not generally possible; in this case it is necessary to resort to comparative validations, and to devote the maximum attention to general validation initiatives like TEAM.

A specific attention to general validation benchmarks, and particularly to wide-coverage, worldwide initiatives like TEAM, appears anyway advisable in general, for any kind of application, since, even when available, experimental measurements do not always possess the required requisites of accuracy and completeness, and codes can also be of interest for applications outside the main production line of the Firm, thus requiring a more general validation..

Problems already released and covering aspects of Electromagnetic Analysis close to those of interest can be directly used both as a benchmark for the selection of the right code and a validation and tuning tool for the existing ones; furthermore, attendance at workshops and proposals of new benchmarks could stimulate a better coverage of other aspects of interest in the near future.

8. Conclusions

In this paper a series of suggestions for a correct and efficient exploitation for industrial applications of Electromagnetic Analysis codes currently available has been presented.

Their use in an industrial environment can be extremely profitable, as already proven by a number of successful cases, but also requires a careful medium term planning and a significant investment in training, that should be also studied in detail.

At the present level of development, this class of tools offers capabilities that, in the opinion of the author, can present a very significant industrial interest for the design of a variety of electromagnetic devices, both advanced and traditional; it appears than likely that this class of tools will reach rather rapidly a much bigger diffusion in industries and design centres than the relatively initial one presently displayed.

Research and development activity for increases of coverage and improvement of user friendliness and postprocessing capabilities is extremely active, and promise, together with the recent advances both in performances and price/performance ratio of workstations, very significant developments of industrial interest in the near future.

If a potential interest in this kind of tools is envisaged, it is then advisable, because of time required to perform and complete a fruitful integration, to start as early as possible a close acquaintance with the area.

9. Acknowledgements

Most of the views presented in this paper are the result of many years of research activity in the "Laboratorio di Progettazione Assistita da Calcolatore di Dispositivi Elettromagnetici" or briefly PACDE Lab, of the Electrical Engineering Department of the University of Genova.

The author is consequently deeply indebted with a large number of persons that have worked with him over these years, and particularly with Paola Girdinio, Paolo Molfino, Maurizio Repetto and Alessandro Viviani, with which it has been established a long and fruitful scientific relationship.

10. References

[1] Turner, L. (Ed.), (1988) "International Workshops for the comparison of eddy current codes" COMPEL, Vol. 7, n. 1 and 2.

[2] Turner, L. (Ed.), (1990) "International Workshops for the comparison of eddy current codes - Round 2" COMPEL, Vol. 9, n. 3.

[3] Nakata, T. (Ed.), (1990) "3-D Electromagnetic Field Analysis" COMPEL, Vol. 9, Supplement A.

ELECTROMAGNETIC DESIGN SOLUTIONS

Jean-Claude SABONNADIERE, FIEEE, FIEE
Professor
Laboratoire d'Electrotechnique de Grenoblen URA-355 CNRS
BP.46 - 38402 Saint-Martin d'Hères, Cédex
FRANCE

ABSTRACT : The use of electromagnetic computer codes in industrial environment is first of all dedicated to the design of electric devices. It is also a good way of understanding the phenomena which are the basis of the operation of the device : fields, fluxes... The paper will focuse the analysis of the aspects of computer codes useful for design and optimization of electric machines and systems.

1. INTRODUCTION

The use of computer codes in Electromagnetic devices engineering becomes today very popular among engineers and scientists. Field computation allows the engineer or the researcher to a better understanding of the underlying phenomena involved in electric or magnetic devices under design. The display of field maps or of colour pictures which show the various levels of a significant physical quantity are very useful to guide the designer and show the impact of various parameters on the basic phenomena.

Moreover, the field computation is today a step in the design process which is a long and difficult task which needs knowledge, creativity and understanding of the electromagnetic laws, the technical requirements and the standards in use in the area of electrical engineering. The general design cycle can be summarized by the flow diagram of figure 1. Starting from the specification, the designer built a preliminary design based on his experience, some scaling rules or sometimes in new project on the rule of thumb. The project can then be improved after the computation of its main output performances and modification of design parameters when the performances checked against the specifications do not fit the latter. After several improvements it happens that the main output parameters are good and the detailed project can be achieved.

The electromagnetic computation codes is involved in the step called analysis and computation which provides the output performances for the test and evaluation step. The code must allows the designer to know, from the current value of each design parameter, the value of all the performance parameters which are involved in the set of specifications. The designer is also interested to get the value of the variation of some performances as function of the design parameters. The possibility to display these values and also some diagrams, plots, colour pictures which illustrate the operation of the devices for the design under consideration is a good help for the improvement of the design. The input of the new set of parameters must be easy in order to get the new values after a short run of the code.

Y. R. Crutzen et al. (eds.), Industrial Application of Electromagnetic Computer Codes, 121–136.
© 1990 ECSC, EEC, EAEC, Brussels and Luxembourg.

A computer code well suited to the design activity must allows all these possibilities associated with intrinsic qualities of a good package : robustness, ergonomy, versatility. We shall analyze in the next paragraphs the detail of the capabilities of a computer code well adapted to the design of electromagnetic devices.

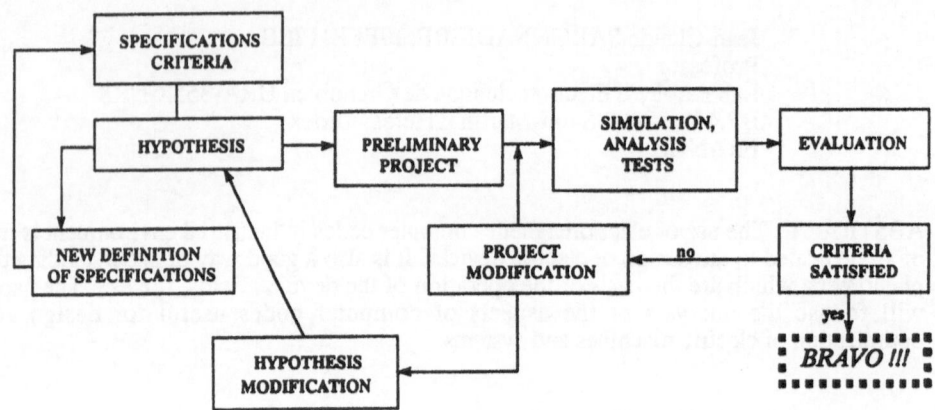

Fig1: ITERATIVE DESIGN

2. ELECTROMAGNETIC MODELS

The design of an electromagnetic device may imply the computation of various characteristics in various electric and magnetic situation. As a matter of fact, the design of a transformer implies the solution of a magnetostatic problem in order predetermine the leakage flux, the electrostatic analysis of the high voltage windings, a magnetodynamic computation of eddy currents for the predetermination of losses, a transient magnetic solution for the analysis of fast transient operation. Moreover, a good design solution needs the knowledge of the thermal behaviour under various operations and an evaluation of mechanical stresses under short-circuit conditions. It appears then clearly that a computer code well adapted to the design of such electric units must be able to provide the capability of all these computations.

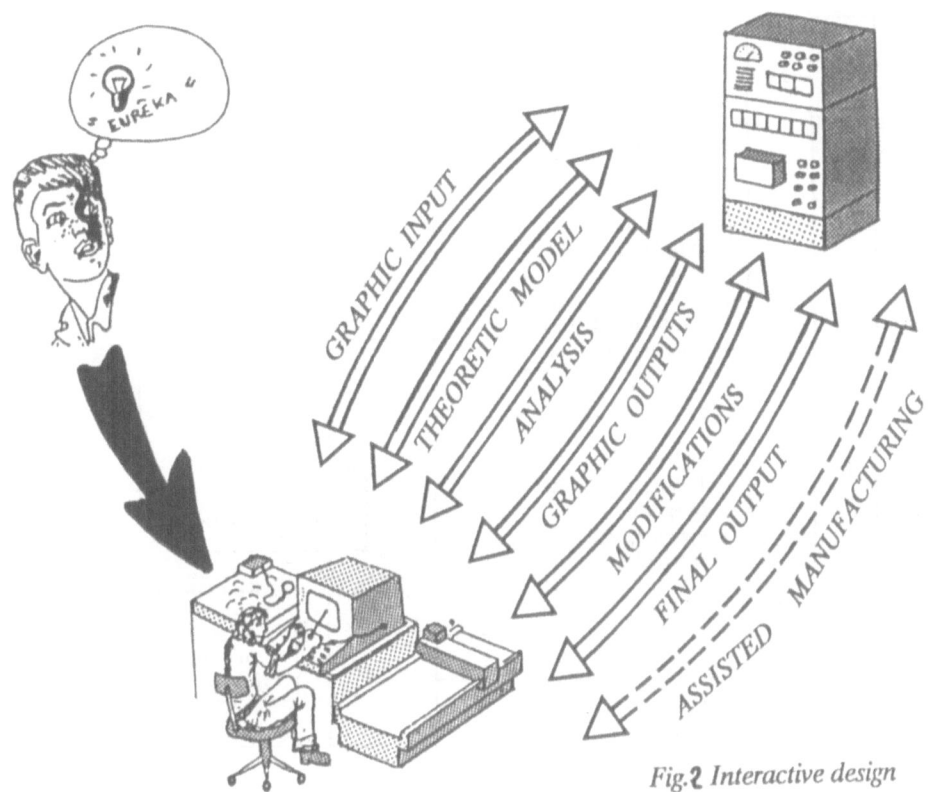

Fig.$\mathbf{2}$ *Interactive design*

Nevertheless, it may be possible that for some reasons it is not possible to find in a single code all those facilities, but, in this case the code must make easy the coupling of its files with the input or output files able to solve the problems which can't be solved by the electromagnetic code. We shall detail this point further in the package architecture. For example if we look at the FLUX 2D software it is noticeable that the global package is made of ten different modules : Electrostatic, Magnetostatic, Electrodynamic, magnetodynamic; electrocinetic, transient magnetic, steady state and transient thermal, magneto-thermal, electrothermal. These scientific modules may be runned on a domain plane or axisymmetric with all adapted boundary conditions. It is thus easy for a designer to make a magnetostatic analysis then to check the design against the eddy current losses by a magnetodynamic analysis and perform a thermal analysis on the basis of the computed losses. At each step the user can define the materials affected to each region and the boundary conditions specific to each problem. The preprocessor and the post processor being the same for each problem, with only an adaptation to specific output data, it is easy for a user to make his design interactively.

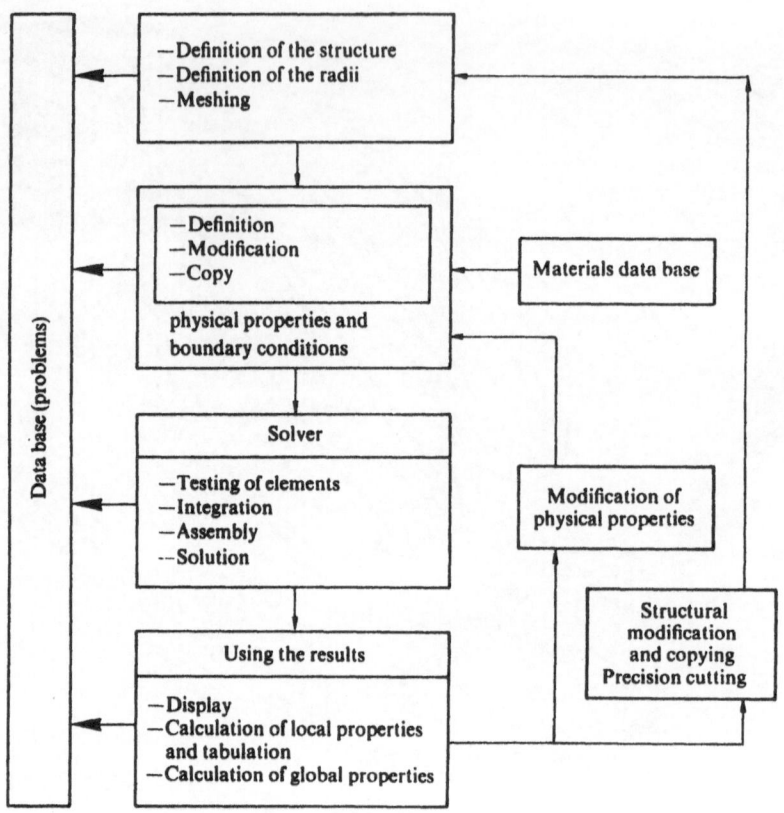

Fig. 3   *Organization of FLUX 2D.*

All these models can be operated on a 2D assumption or more generally on a 3D basis. The problem of making a 2D analysis or a 3D one for the design wil be discussed after the analysis of these computations.

3.   2D COMPUTATION IN ELECTROMAGNETICS

3.1   Principles

When we use computer software in order to analyse an electromagnetic device, we think generally that fields are perfectly three-dimensionnal.

However when we look deeper in the problem it appears that if the problems seems to have a 3D structure, generally there is a sort of anisotropy in the architecture (electric machines are generally long, contacts in the apparatus are cylindrical) or in the physical aspect of the device (laminated magnetic circuit keeps the flux in the plane of the laminations, winding elements are circular and thus axisymmetric, eddy currents are perpendicular to the field...) which penalizes one dimension with respect to the others.

In this situation the importance of the phenomenon in this direction becomes of the second order compared with its effect in the plane of the other directions. This analysis is very frequent in electrical engineering, even if sometimes some small disturbance in the last direction makes the analysis more difficult, in order keep a good accuracy in the computation of the global behaviour of the system.

In fact we can consider four kinds of problems which may be studied by a bidimensionnal analysis : axisymmetric structures in which the third dimension is the azimutal angle, infinitely long systems in which the phenomenon is analyzed in the x-y plane and is considered as invariant in the z directions ; for both kinds of devices we can analyse either scalar potential problems or vector potential derived fields.

3.2   Plane 2D Systems

This subparagraph applies to devices which can be assumed invariants in the z direction and in which the field distribution is analyzed in the x-y plan, cross section of the device. Two kind of problems can be stipulated according to the physical nature of the problem.

a)   Scalar potential problems

These problems occur when the Maxwells equations are Curl free i.e if we write $\nabla \times H = o$ or $\nabla \times E = o$. In this case there exists a scalar function which is either the scalar magnetic potential $\emptyset$ or the electric potential V such as : $H = \nabla \emptyset$, $E = -\nabla V$.

Then, the divergence theorems lead to the final equation :

$$\nabla . (-\mu \nabla \emptyset) \tag{1}$$

$$\nabla . (-\varepsilon \nabla V) = \rho \qquad (\rho: \text{charge}$$

density)

These equations expressed in the x-y plane become :

$$\partial/\partial x(\mu \, \partial \emptyset/\partial x) + \partial/\partial y \, (\mu \, \partial \emptyset/\partial y) = o$$

or :
$$\tag{2}$$

$$\partial/\partial x(\varepsilon \, \partial V/\partial x) + \partial/\partial y \, (\varepsilon \, \partial \emptyset/\partial y) = -\rho$$

in which the parameters μ and ε depend generally on the field strenght which makes the problem non linear ($\mu = f(/B/^2)$, $\varepsilon = f(/E/^2)$).

The boundary conditions may be Dirichlet conditions (known potential), Neuman condition (along a field line) or some specific condition like periodic or floating potential conditions. It must be pointed out that isopotential are perpendicular to field lines.

b)   *Vector potential problems*

This kind of problems occurs in magnetic systems when the right hand of the curl equation is not equal to zero (a current density exists).

Fortunately in that case the field is divergence free. If the flux density B is divergence free, that means that there is a vector A such as $B = \nabla xA$ (because a curl vector is always divergence free) which is named the vector potential. As the field in the x-y plane and is invariant in the z direction the vector A has only a non zero component is the z direction and thus A can be represented by it z composant A such as :

$$\nabla x(\nabla xA) = J \tag{3}$$

where J is the current density. The projection along the z direction of equation (3) leads to:

$$\partial/\partial x(v\, \partial A/\partial x) + \partial/\partial y\, (v\, \partial A/\partial y) = -J \tag{4}$$

in which the reluctivity $v = (|B| / |H|)^{-1}$ depends on the flux density strenght $|B|$ making the problem non linear.

As previously mentionned, boundary conditions are of Dirichlet, Neuman, periodic of floating type. It must be noticed here that on the contrary of the scalar potential problem that Neuman conditions applies to boundaries which are perpendicular to the field lines (symetry conditions).

3.3   Axisymmetric structure (cylindrical coordinates)

In this kind of device all physical quantities are constant on a circle which has his center on the z axis, that means that field and flux density are constant for all values of the azimutal angle.

a)   *Scalar potential*

Let us consider a scalar potential : or B the fields E or H are the gradient of this potential and must fulfill a divergence equation :

$$\nabla.(\mu\, \nabla\emptyset) = o$$

$$\nabla.(\varepsilon\, \nabla V) = \rho$$

In cylindrical coordinates these equations becomes :

$$1/r\, \partial/\partial r\, (\mu\, r\, \partial\emptyset/\partial r) + \partial/\partial z\, (\mu\, \partial\emptyset/\partial z) = o$$

$$1/r\, \partial/\partial r\, (r\, \varepsilon\, \partial V/\partial r) + \partial/\partial z\, (\varepsilon\, \partial V/\partial z) = -\rho \tag{6}$$

The boundary conditions are the same that the conditions which accur in cartesian coordinates.

b)   *Vector potential problems*

In this kind of problems the equations :

$$\nabla x(v\, \nabla xA) = J$$

becomes taking into account the fact that the component of A along r and z are either zero or constant :

$$\partial/\partial z(v\, (-\partial A/\partial z)) - \partial/\partial r\, (v/r\, \partial/\partial r\, (r\, A_\varphi)) = J \tag{7}$$

which may be rewritten :

$$\partial/\partial r \, (v/r \, \partial/\partial r \, (r \, A_\varphi)) + \partial/\partial z \, (v/r \, \partial A_\varphi/\partial z) = J\varphi \qquad (8)$$

the boundary conditions are similar to these exhibited here up with in addition the Dirichlet conditions $A_\varphi = o$ along the z-axis (r=o) because A_φ must be zero at infinity and by continuity everywhere along the z-axis.

All these problems may be integrated in a general software allowing the solution of various physical problems encounted in physics or in electrical engineering.

For instance in the FLUX 2D software is a comprehensive package which allows the solution of the a large variety of problems :

Electrostatic Electrocinetics	scalr potential
Magnetostatic Magnetodynamic Transient magnetic	vector potential
Electrodynamic	field
Thermal problem (steadystate) Thermal problem (transient) Magneto-thermal (coupled) Electro-thermal (coupled)	temperature

These equations can be associated with various boundary conditions previously expressed: Dirichlet, Neuman, Cyclic, anticyclic, floating Translation, convection, radiation ...

4. THREE DIMENSIONAL COMPUTATION IN ELECTROMAGNETISM

The 3D computation is very different of the previous 2D one because its implies the complete description of the device with its hidden regions and all the significant details. On the mathematical point of view the difference is also noticeable for vector potential problems for which each node bears 3 unknowns components and by the fact that the boundary conditions must be expressed on surfaces instead of lines.

4.1 Principles

For three dimensional structures the scalar potential formulation must be used when it is possible for multiple reasons. Where such a formulation does not exist it is always important to try to built an hybrid formulation in which the vector calculation is reduced.

4.1.1 *Scalar potential*

The electric scalar potential is used for problems in electrostatics and for the solution of current density distribution in conductor. In both kind of electric field analysis the equation to be solved is :

$$\nabla.(\gamma \nabla V) = 0 \tag{9}$$

in which γ is the permittivity (respectively the conductivity of the media.
The boundary equations are generally Dirichlet or zero Neuman conditions.

For the analysis of magnetic field we can use a scalar total magnetic potential, if the field source is a permanent magnet or a boundary condition but not a current density. When permanent magnets are included in the domaine the potential φ must fulfill the equation :

$$\nabla.(\mu \nabla \emptyset) = \nabla.Br \tag{10}$$

where Br is the remanent flux density (when the field H is equal to zero) and μ the magnetic permeability of the media.
The boundary conditions are generally Neuman or Dirichlet conditions.

If there is a non zero current density in the domain it is possible to define a formulation with a reduced scalar potential. This formulation consists in breaking down the magnetic field H into two components : an excitation field He created by the current density J and a component Hm which expresses the magnetic material reactions and is curl free. The field He can be computed by the Biot and Savart law and Hm will be represented by a reduced potential ψ such as Hm=-Vψ and must fulfill the equation :

$$\nabla.(\mu \nabla \psi) = \nabla.(\mu He + Br) \tag{11}$$

with Neuman and/or Dirichlet conditions.

4.1.2 *Vector potential*

This formulation can be used when we have current density and the divergence equation on the flux density B leads to introduce the vector potential A such as B= ∇xA. The equation on magnetic field H and the relation betwen B and H leads to :

$$\nabla.(\nu \ \nabla xA) = J - \nabla xHc$$

where Hc is the coercitive field in permanent magnets.

$$(12)$$

A problem which arises in this formulation is the gauge problem because A is not unique if no supplementary condition is defined. Generally the Coulomb gauge ∇.A=o provides the unicity of A. Various boundary condition like Axn=o, Hxn=o may be applied with also of course of condition A=o at infinity.

Some other formulations exist like hybrid formulation called A-φ in which there is a vector potential A in the regions in which there is a current density and the scalar potential φ in current free regions. This formulation which is expensive in computer time is not very adapted to magnetostatic problems but will be used for eddy currents solutions.

5 .   COMPARISONS OF 2D AND 3D SOFTWARES

5.1   Advantages of 2D softwares

When we try to compare the advantages and drawbacks of 2D and 3D analysis we must take into account the various parameters of the design or the analysis of a device : Flexibility and amount of work required to model the device, computer time consumption, possibility of parameter analysis and optimization, capability of interpretation of the results of computation, accuracy of the results, etc... If we look deeper in the qualities of 2D package for these various parameters we can state that 2D analysis is :

* Flexible
> That means the time required for the description of the device, the finite element meshing, the assignation of material in each internal regions and boundary conditions on the limits of the domain is moderate and never exceed half a day or more even for complex structures.

* Moderate in Computer time Consumption
> Generally a 2D problem do not exceed 10.000 nodes and has an average of 2.000 nodes for second order finite elements grids. This size of problems leads to computer time on mini-computer or on workstation which are in the range of one second to several minutes depending of the size of the problem and the power of the computer.

* Allows parametric analysis
> If the CPU time required for the solution of one problem in small and if the software allows the possibility of definition of parameters it is possible to make a sensitivity analysis of a performance versus various parameters and get a good design.

* And even optimization

As the sensitivity analysis is the first step towards optimization the possibility of describing a parametrized structure and of launching a sensitivity analysis provides the capability of making complete optimization of a device.

* Provides an easy interpretation of the results

A 2D problem can be easily mapped on a plane and displayed on a screen. The usual method of appreciation of the results isoflux lines, vectors, colors shading can be very easily interpreted by the physicist and the engineer.

* Provides an accuracy consistent with the physical assumptions of the problems

In fact, as far as, the assumption of the 2D problems is fulfilled the accuracy is very good or in the range of expected value. We begin to meet a problem when for instance some thing happens in the third dimension and makes the assumption hazardous. In this kind of situation, it is often possible to bring solutions like perform an analysis in various cross sections or to compute the fields in two orthogonal planes in order to make interpolations or extrapolations. Nevertheless some problems are actually three dimensonals without any favoured direction for which two dimension analysis fails to proved good results, however this kind of problems are not very frequent.

5.2 Advantages and drawbacks of 3D analysis

The main advantage of the 3D analysis is its exactness of the theoretical foundation of the problem. When a 3D analysis is made one can be sure that the Maxwell's equation are rigorously respected in theory.
Now, one question arises to know if it useful to respect exactly the 3D expression of Maxwell's equations if, on the other hand the finite element discretization introduces discretization errors and/or round off error during the calculation ? But, first at all let us examine the key points of a 3D analysis.

a) Data description

The description of a device in 3D is a work long and cumbersome which needs a great amount of human time specially if the device is made of parts of complex geometric shape. For instance the figure . represents the part of a switchgear in which the electric ar is blown up. Its description on the PHI3D software which is very convivial needs several hours, the problem being to detect the level of detail which must be described.

b) Mesh generation

The finite elements or boundary elements mesh generaltion is a very difficult problem, particularly in electromagnetism where very small geometric air gaps plays an important part on the magnetic behaviour of the device.
At our knowledge there is presently no automatic mesh generators which provides for every problem a good finite element grid. It appears that in order to obtain a correct finite element grid, the users must steer the mesh generator or correct it results. This work is also long and cumbersome and needs very powerful display functions in order to look into great details the generated mesh.

c) Finite element solution

The third step in a 3D analysis is the solution of the (eventually non linear) system of equations stemming from the finite element discretization. The system of equations is solved when the problem is non linear by the Newton-Raphson method which lead to the solution of linear equations with a sparse band matrix in the finite element method.
Looking into the finite element method the average problems gives a system of 20.000 to 30.000 nodes and thus equations.

If the problem is solved by a boundary integral method the number of equations is only 2.000 or 3.000 but with a full matrix that is probably worse than the previous one.
The solution of this kind of problems needs a supercomputer or becomes very long (several hours) on a standard mini-computer or workstation. The large computer time consumption make any sensitivity analysis impossible because very costly or too long and thus forbides any actual optimization.

d) Post processing

The interpretation of a field of vectors on a display screen needs a long experience of this kind of problem. All the possible representations (colors, vectors, surfaces) are generally not easy to understand clearly, thus the best way is to get cross-sections and examine the phenomena in these sections. However, this is not obvious because for instance the vector potential lines have no meaning when they are the plane section of a 3D field.
This discussion shows that for specific analysis the use of 3D packages may be essential to be sure of the goodness or the accuracy of the results. However, the heaviness of handling these software makes their use difficult for design purpose or optimization processes. In our mind, they must be kept to check the final result of a design or the performances of an optimized product.

6. ARCHITECTURE OF COMPUTER AIDED DESIGN ELECTROMAGNETIC CODES

In order to insure all the qualities required for a code to be adapted to the design procedure some capabilities must exist in order to ease the modification of the characteristic data of the device under design. This architecture is based on the concept of data bases adapted to the project. A data base is a collection of informations which is representative of an object and allows to describe all the various parts of this device and assign to each part a material which is in turn extracted from an other data base.

Thus the general architecture of an electromagnetic computer code is according to the diagram of figure (4), made of some general modules similar to those fore mentionned and which are connected through a data-base to internal or external modules for geometric design, mesh generation post-processing or mechanical and/or thermal analysis computer codes.

In fact, inside a company the electromagnetic design office is often one link in the chain of product design. Above we find the geometric design which is performed on a CAD product like autocad, cadam, euclid... and terminates on the geometric definition of various parts of the device with the specifications of the material in each part. Below we find the softwares for thermal and mechanical design, the code for the definition of the plastic rheology of sheaths and packaging materials... and finally the computer aided manufacturing codes.

The possibility of safe communications of the C.A.D. electromagnetic code and the other softwares of the chain is need for modern applications of these packages.

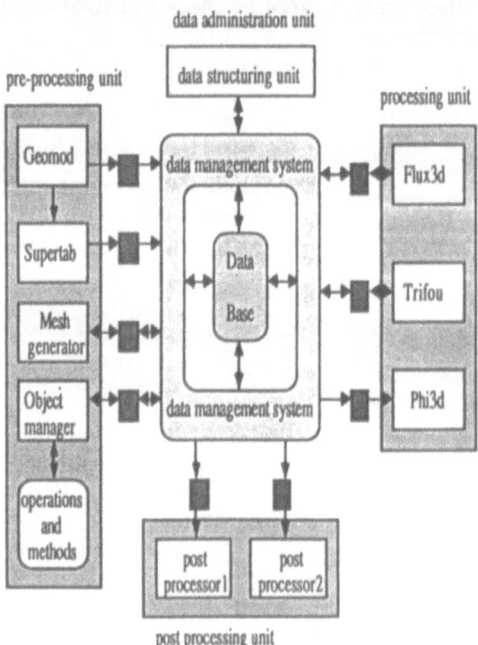

Fig. 4 Database environment.

7. OPTIMIZATION OF ELECTROMAGNETIC DEVICES

The design solution implies implicitely one step which is the optimization of the device versus several design parameters. Recent original works done by Coulomb [ ] allows an automatic procedure of optimization based on the sensitivity analysis. We shall present in this section two examples of optimization in electromagnetics.

The first one will consider the optimization of the shape of a permanent magnet in order to fulfill fieds requirements. The second considers the optimization of the current density in the winding and the reluctivity of the magnetic circuit of the electromagnet of a contactor.

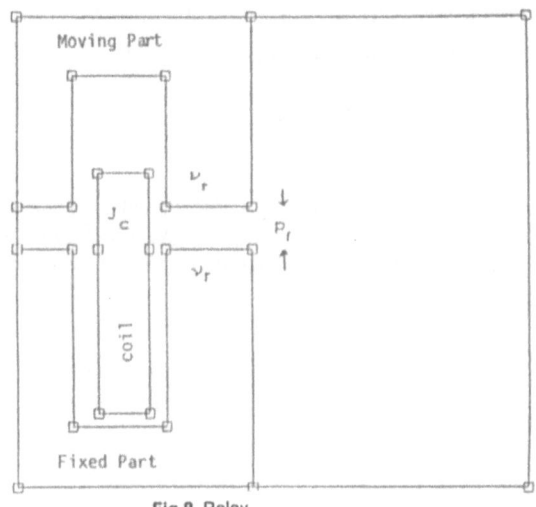

Fig.8 Relay.

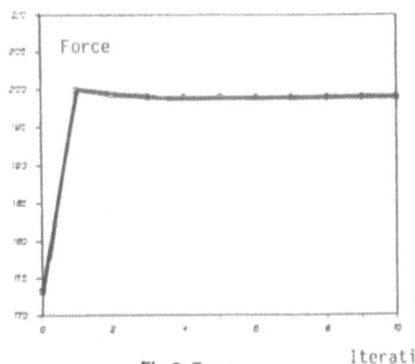

Fig.9 Force.

7.1 Optimization of the shape of a permanent magnet

The problem is to definie the height parameters p_1, p_2, p_3, p_4 of the various parts of a permanent magnet of permeability $\mu r = 1$ and coercitive magnetic field Hc = 2000.000 A/m in order to insure a sinusoïdal given distribution of the flux density along the y-axis at the points Q_1, Q_2, Q_3, Q_4 indicated on fig....

In this problem the objective function will be defined by :

$$Z(p) = \Sigma \ [(\bar{B}_j - B_j)/B_j]^2 \qquad (13)$$

in which $\bar{B}_j$ is the given and B_j the expected value of the component by of the flux density at the point Q_j. It is foreseen to minimize this objective function.

As in any problem the parameters p_i are bounded by p_u and p_e which define the constraints

$$g_i \ (p_i) = p_i{}^2 + p_i \ (p_u + p_l) - p_u \ p_i > 0$$

for each parameter p_1 to p_4.

The constrainted problem is transformed into an unconstrained one by use of a special function calld penalty function. The fig.6 shows the final result when the difference between the imposed values and the actual computed value is less than 1 % on the B_y component.

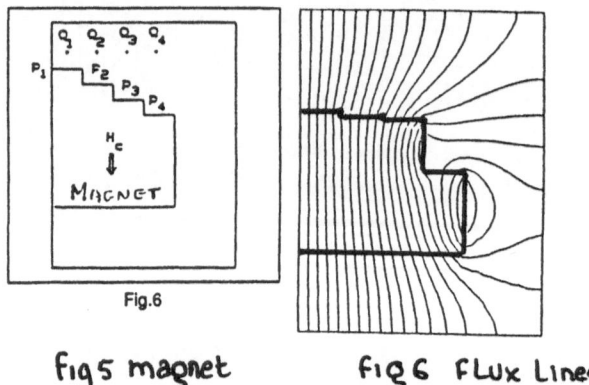

Fig.6

fig 5 magnet   fig 6 flux Lines

7.2 Optimization of a relay

This example will illustrate the search of optimal value of electric and geometric parameters. The model of the relay is shown on fig... The two parameters used in the design are the air-gap p_f and the current density J_c while keeping the force between armatures at a given value (for instance 2N), the reluctivity v_r being constant ($v_r = 4.0 \ 10^{-3}$).

The constraints will be the upper (p_u) and lower (p_l) bounds of J_c and p_f defined by the equations (1) in which i takes two different values, and the equality constraint on the force $F = F_o$. This latter constraint can be transformed into an inequality constraint by use of a tolerance ΔF :

$$g3\ (p_1, p_2) = (\Delta F)^2 - (F - F_o)^2 > 0 \qquad\qquad (14)$$

After elimination of the constraints by a penalty function the minimization is made by a gradient method. The result gives a value of $p_f = 4.047$ mm, $j_c = 7.254$ A/mm^2 for a force $F = 1.99$ N. The figure 9 shows the variation of the parameters during the computation.

We have presented here some example of optimum design of electromagnetic devices based on the finite element method. We tried to illustrate the importance of the definition of design constraints in order to solve an actual engineering problem.

In all cases optimum design remains of time consuming problem but a systematic approach, even heavy, is perhaps better that an empiric random approach of the design.

8. CONCLUSION

We tried all along this paper to show that if 3D calculations are sometimes essential to be sure to get a good answer to the design problem, this kind of computation is always heavy to handle and not very convenient to integret.
On the other hand 2D computation are theoretically les accurate because they do not take the third dimension into account, but on reasonable assumptions they can become very useful because they are flexible, robust and easy to run even iteratively for optimization

procedures. In the author's opinion it is generally adivised to run one design problem on a 2D package and analyse all the aspects in both directions in order to acquire a good knowledge of the problem then, if necessary, launch a sensitivity analysis or an optimization algorithm. These computation allows the designer to know approximatively the value of the design parameters of this device. Afterwards, the performance reached can be computed and checked by few 3D computations on the optimal parameters, and output beautiful colour picture on the flux density or any other physical quantities.
This methodology is able to optimize the designer work and save large amount of computer time.

REFERENCES

[1] B. ANCELLE, J.L. COULOMB, B. MOREL, E. BELBEL
Implementation of a computer aided design system for electromagnets in an industrial environment. IEEE Trans-Mag, Vol. May 16, n°5 , Sept.1980

[2] G. MEUNIER, J.L. COULOMB, J.C. SABONNADIERE
2D and 3D finite element modelling of small electrical machines. IMACS Conference, Liège (Belgium) 1984

[3] J.L. COULOMB, G. MEUNIER
Finite element analysis of a self starting synchronous micromotor. IEEE Trans-Mag. Vol 17, n°6, 1981

[4] S. GITOSUSASTRO, J.L. COULOMB, J.C. SABONNADIERE
Performances derivatives calculations and optimization process. IEEE Trans-Mag, Vol. July 1989

[5] J.L. COULOMB
A methodology for the determination of global electromechanical quantities from finite analysis and its application to the evaluation of magnetic forces, torques and stifness. COMPUMAG Conf.Genova IEEE Trans-Mag., Nov.1983

[6] J.L. COULOMB and G. MEUNIER
"Finite element implementation of virtual work principle for magnetic or electric force and torque computation. INTERMAG Conf., 10-13 April 1984 Hamburg, IEEE Trans-Mag., Sept. 1984

[7] J.L. COULOMB and J.C. SABONNADIERE
Finite element and CA. Springer Verlag, New-York, 1987

[8] B. MOREL, M. ANDRE and J.C.SABONNADIERE
A parametric preprocessor for FLUX 2D. IEEE Workshop on Electromagnetic field computation Schenectady 1986.

ADVANCED 3D ELECTROMAGNETIC FORMULATIONS
AND APPLICATIONS

J.C. VERITE

Electricite de France
Direction des Etudes et Recherches
Clamart   FRANCE

1. A formulation using the magnetic field for the 3D Eddy current problem (A. Bossavit, J.C. Vérité) ([1], [2], [3], [4]).

1.1. DEFINITION OF THE PROBLEM

Let E be the physical space and consider two bounded subdomains Ω and Ω_0 of E (Fig. 1), with $\Omega_c = E - \Omega$, and let Γ be the boundary of Ω.

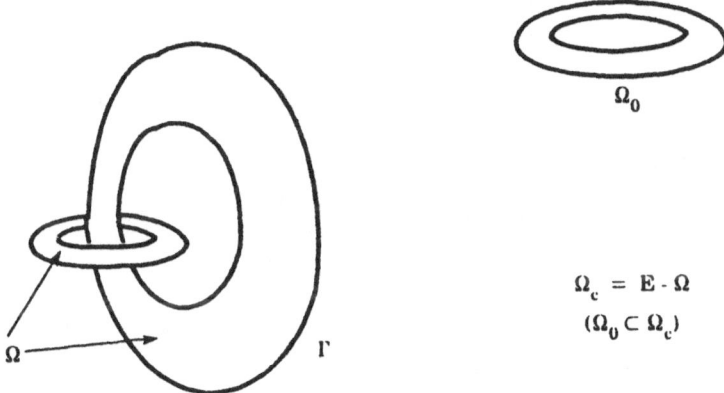

$$\Omega_c = E - \Omega$$
$$(\Omega_0 \subset \Omega_c)$$

Fig. 1 - Space decomposition

Let $j_0(x, t)$ be a known function (with $\operatorname{div} j_0 = 0$) defined on Ω_0 and let $\rho(x) (> 0)$ and $\mu(x) (\geq \mu_0 = 4\pi\, 10^{-7})$ be two coefficients defined on Ω.

From a physical point of view, Ω is a passive set of conductors with resistivity ρ and permeability μ and Ω_0 is an active coil with an imposed current j_0.

Y. R. Crutzen et al. (eds.), Industrial Application of Electromagnetic Computer Codes, 137–160.
© 1990 *ECSC, EEC, EAEC, Brussels and Luxembourg.*

We want to calculate the current density j in Ω and the magnetic field h_t at any time t:

  - either as an "initial" or "Cauchy" problem knowing the field at instant 0,

  - or as a periodic problem, assuming that $j_0(x, t)$ is a periodic function of time.

1.2. DECOMPOSITION OF THE TOTAL MAGNETIC FIELD

Let h_0 be the magnetic field due to j_0 on $Ω_0$ supposed to be alone in E (Ω does not exist).

We can explicitly calculate h_0 using the Biot and Savart's law.

In the situation of Fig. 1 the total field h_t differs from h_0 if there are induced currents in Ω and/or the permeability of Ω is not equal to $Ω_0$.

We can then write that:

$$h_t = h + h_0.$$

where h, the unknown variable of our problem, is the "reaction field" due to the presence of Ω.

Remark:

In this problem we consider that j_0 is the current flowing in $Ω_0$ when Ω exists. So the starting situation without Ω is a little bit artificial since we should have $j'_0 \neq j_0$ in that case.

1.3. THE VARIATIONAL FORMULATION

Physically we need

$$\int_E \left| h(x, t) \right|^2 dx < \infty \quad \text{and} \quad \int_E \left| \operatorname{curl} h(x, t) \right|^2 dx < \infty.$$

We can then define the following Hilbert space H for h:

$$H = \left\{ h \in L^2(E), \operatorname{curl} h \in L^2(E), \operatorname{curl} h = 0 \text{ in } Ω_c \right\}$$

with the scalar product:

$$(h, h') = \int_E h\, h' + \int_E \operatorname{curl} h \cdot \operatorname{curl} h'$$

Remark:

The previous definition imposes the continuity of the tangential component of h through a surface, but the normal component can be discontinuous.

Let us start from the Faraday's law:

$$\mu \frac{\partial}{\partial t}(h + h_0) + \operatorname{curl} e = 0$$

if we multiply by any $h' \in H$ and integrate over E we get:

$$\frac{d}{dt}\int_E \mu(h + h_0)h' + \int_E \operatorname{curl} e \cdot h' = \frac{d}{dt}\int_E \mu(h + h_0)h' + \int_E e \cdot \operatorname{curl} h' = 0$$

using Ohm's law ($e = \rho \operatorname{curl} h$) and the definition of H ($\operatorname{curl} h' = 0$ in Ω_c) we finally get:

$$\boxed{\frac{d}{dt}\int_E \mu(h + h_0) \cdot h' + \int_\Omega \rho \operatorname{curl} h \operatorname{curl} h' = 0 \quad \forall h' \in H} \qquad (1)$$

with some regularity conditions on h_0, it can be proved that the Cauchy or the periodic problem associated to (1) has a unique solution.

On the other hand it can also be proved (cf. [2]) that the unique solution h solving (1) is such that $h + h_0$ is the total magnetic field of the physical problem.

1.4. REPRESENTATION OF THE FIELD OUTSIDE CONDUCTING REGIONS. INTRODUCTION OF THE "POINCARE - STEKLOV OPERATOR"

Outside Ω we have $\operatorname{curl} h = 0$. We can then write $h = \operatorname{grad} \Phi$ in Ω_c, where Φ is a magnetic scalar potential.

Let φ be the value of Φ over Γ. We have also $\varphi = 0$ at infinity for uniqueness and $\operatorname{div} h = 0$ in Ω_c.

The variable Φ in Ω_c is thus the solution of the following problem:

$$\Delta\varphi = 0 \text{ in } \Omega_c$$
$$\left\{ \begin{array}{c} \varphi = 0 \text{ at infinity} \\ \varphi = \varphi_\Gamma \text{ on } \Gamma \end{array} \right.$$

It is clear that φ and then $\partial\varphi/\partial n$ over Γ, depends only on the value φ_Γ of φ on Γ. It is thus possible to write:

$$\frac{\partial\varphi}{\partial n_\Gamma} = \mathbf{R}\,\varphi_\Gamma$$

where $\mathbf{R}$ is an integral operator, the "Poincaré Steklov operator", linking the values of $\partial\varphi/\partial n$ on a point of Γ to the values of φ everywhere on Γ.

In (1) we have a term:

$$\frac{d}{dt} \int_{\Omega_c} \mu_0 h \cdot h'$$

that we can transform in the following way:

with $h = \text{grad } \varphi$ and $h' = \text{grad } \Psi \in \Phi$

$$\Phi = \left\{ \varphi, \varphi \in L^2(\Omega_c), \text{grad}\varphi \in L^2(\Omega_c) \right\}$$

We get:

$$\frac{d}{dt} \int_{\Omega_c} \mu_0 \text{grad } \varphi \cdot \text{grad } \Psi = \frac{d}{dt} \int_\Gamma \mu_0 \frac{\partial\varphi}{\partial n} \Psi = \frac{d}{dt} \int_\Gamma \mu_0 \mathbf{R}\varphi\psi.$$

Using the Green's theorem (since we have $\Delta\varphi = 0$ in Ω_c) we have also:

$$\frac{d}{dt} \int_{\Omega_c} \mu_0 h_0 h' = \frac{d}{dt} \int_{\Omega_c} \mu_0 h_0 \text{grad } \psi = -\frac{d}{dt} \int_{\Omega_c} \mu_0 \psi \text{ div } h_0 + \frac{d}{dt} \int_\Gamma \mu_0 \cdot h_0 \cdot n \psi$$

$$= \frac{d}{dt} \int_\Gamma \mu_0 \cdot h_0 n \psi \text{ since div } h_0 = 0 \text{ in } \Omega_c.$$

The variational formulation that we finally obtain is the following:

Find $h \in H$ and $\varphi \in \Phi$ such that $\forall\, h' \in H$ and $\Psi \in \Phi$:

$$\frac{d}{dt}\int_{\Gamma}\mu_0\,\mathbf{R}\varphi\psi + \frac{d}{dt}\int_{\Omega}\mu\,h\,h' + \int_{\Omega}\rho\,\mathrm{curl}\,h\,\mathrm{curl}\,h' = -\frac{d}{dt}\int_{\Gamma}h_0\cdot n\,\psi - \frac{d}{dt}\int_{\Gamma}\mu\,h_0\,h' \quad (2)$$

2.   The discretized formulation (A. Bossavit, J.C. Vérité)

The previous variational formulation involves two variables, h in Ω and Ψ on Γ. We use finite elements for h in Ω and also for Ψ on Γ.

To do that Ω is decomposed in tetrahedra which give plane triangular faces for Γ. φ is then classically decomposed using piecewise linear elements on these triangular faces.

The originality of the method we present appears in the use of "edge elements" for h in Ω ([4]) and in the use of the operator $\mathbf{R}$ and the way we get a discretized expression for it using an integral method ([3]).

2.1.   USE OF WHITNEY'S ELEMENTS - EDGE ELEMENTS

A. Bossavit has shown, for example in [4], that Maxwell's equations could be best expressed in terms of differential forms than in terms of vector fields. Recall that a differential form is an alternating multilinear application from R^p to R. For example h can be seen as a 1-differential form, for which the integration over a 1-dimensional domain makes sense. We all know that what has a physical meaning for h is precisely the circulation along a path. As for j, it can be seen as a 2-form. We have thus at our disposal a mathematical tool very well adapted to the physics.

Whitney forms ([5]) are a family of differential forms on a simplicial mesh (tetrahedra), such that a Whitney 1-form play the role of finite elements for h for example, the degrees of freedom of which being associated with edges of the mesh. A Whitney 1-form is what we call an "edge-element", discretizing a differential 1-form, h. An important practical consequence of that correspondance is that using such elements we impose the continuity of the tangential component of h, but not that one of the normal component, fitting thus with interface conditions.

A Whitney 0-form can be seen as a finite element with degrees of freedom corresponding to an integration over a domain of dimension 0, i.e. value at one point. This is a classical nodal finite element adequate to discretize a 0-form such as the magnetic scalar potential Ψ.

We have just seen what is a Whitney 1-form for discretizing a 1-form such as h or e. More precisely if i and j are two vertices, λ_i and λ_j the corresponding barycentric coordinates, the function w_{ij} associated to the edge ij is:

$$W_{ij} = \lambda_i \nabla\lambda_j - \lambda_j \nabla\lambda_i .$$

The circulation of such a function is equal to 1 along the edge ij and 0 along any other edge.

We have then $h(x) = \Sigma\, h_{ij}\, w_{ij}(x)$, the coefficient h_{ij} being the circulation of h along ij.

We could define also Whitney 2-forms adapted for discretizing differential 2-forms such as b or j. These are "facets elements", the degrees of freedom of which are the fluxes across facets, the significant quantities for b and j.

Finally in equation (2) we write

$$h(x) = \sum_{ij\,\in\,IE} h_{ij}\, w_{ij}(x) + \sum_{s\,\in\,BN} \varphi_s\, V_s(x)\ \text{ with } V_s = \operatorname{grad}\lambda_s$$

where IE is the set of inner edges and BN the set of boundary nodes.

We then take every w_{ij} and every V_s as test functions.

2.2. A BOUNDARY INTEGRAL METHOD FOR **R**

We have now to find a suitable matrice approximation R for **R**, i.e. a square matrix whose order is the number of boundary nodes.

To do this we can solve the exterior problem in φ using a boundary integral method.

We can use the third Green's formula:

$$\frac{1}{2}\,\varphi(x) + \frac{1}{4\pi}\int_\Gamma \frac{n(y)(x - y)}{|x - y|^3}\,\varphi(y)\,dy = \frac{1}{4\pi}\int_\Gamma \frac{dy}{|x - y|}\,(R\varphi)(y) \qquad (3)$$

We can also use a single layer of charges q:

$$\varphi(x) = \frac{1}{4\pi}\int_\Gamma \frac{q(y)}{|x - y|}\,dy \qquad (4)$$

with:

$$\frac{\partial \varphi}{\partial n}(x) = \frac{q(x)}{2} + \frac{1}{4\pi} \int_{\Gamma} \frac{n(y)(x-y)}{|x-y|^3} q(y)\,dy \qquad (4')$$

$$x \to x_0 \in \Gamma$$

We can then use, either in (3) or in (4) and (4'), a point-matching method or a variational method. The second one gives better results (see [6]). We multiply (3) by φ or (4) by q' and (4') by φ, test functions for q and φ and we integrate over Γ. Using the discretized expression for φ and q we get:

$$(3) \Rightarrow (\frac{1}{2} + H)\varphi = GR\varphi \Rightarrow R = G^{-1}(\frac{1}{2} + H) \qquad (5)$$

(4) and (4') $\Rightarrow \varphi = Gq$

$$R\varphi = q + H^T q \Rightarrow R = (\frac{1}{2} + H^T)G^{-1} \qquad (6)$$

where G and H are matrices.

The matrices R are different in (5) and (6) (they are transposed one each other), because the symmetry of the operator R is lost in the discretization process.

We can take $(R + R^T)/2$.

An other solution would be to use also the Poincaré Steklov operator **R** of the inner region Ω. In that case we would get a symmetrical expression for R (see [3]).

In TRIFOU we have used a variational method for (4) and (4'), i.e. expression (6).

3. An associated 3D formulation using the electric field e as state variable (A. Bossavit, Z. Ren)

In the previous formulation (2) using h as the state variable, the "magnetic formulation", the electric field e can be calculated in the conducting regions by $e = \rho \, \text{curl} \, h$, but it cannot be determined outside the conducting regions. We present now an alternative formulation, the "electric" one, using e as state variable (cf. [7], [4]).

The Ampère's theorem can be expressed variationaly:

$$\int_E h \cdot \text{curl } e' - \int_\Omega \sigma e \cdot e' = \int_{\Omega_0} j_0 \cdot e'$$

with:

$$e' \in E = \left\{ e \in L^2(E), \text{curl } e \in L^2(R^3), \text{div } e = 0 \text{ in } \Omega_c, \int_\Gamma n.e = 0 \right\}$$

$$\Rightarrow \int_E \mu^{-1} b \text{ curl } e' - \int_\Omega \sigma e \cdot e' = \int_{\Omega_0} j_0 e'$$

deriving in time this last equation and using Faraday's law:

$$\frac{\partial b}{\partial t} + \text{curl } e = 0$$

we get:

$$\int_E \mu^{-1} \text{curl } e.\text{curl } e' + \frac{\partial}{\partial t}\int_\Omega \sigma e.e' = -\frac{\partial}{\partial t}\int_{\Omega_0} j_0.e' = -\frac{\partial}{\partial t}\int_{\Omega_c} \text{curl } h_0.e' = \frac{\partial}{\partial t}\int_\Gamma \mu^{-1} e'n \times b_0$$

This formulation involves the whole space E. Once again, as for the magnetic formulation, we want to use a boundary integral method outside the conducting region Ω.

The great difference with the magnetic formulation is that in this case we do not introduce any scalar potential and we solve the problem everywhere in term of e.

Integrating by parts ana taking into account the fact that curl b = 0 in Ω_c we get:

$$\int_E \mu^{-1} \text{curl } e \text{ curl } e' = \int_\Omega \mu^{-1} \text{curl } e \text{ curl } e' + \int_\Gamma \mu^{-1} e'.n \times \text{curl } e$$

We can then, in analogy with the h formulation, introduce an integral operator S such that on the boundary Γ, with n the normal unit vector pointed inward Ω, we have:

$$\int_\Gamma e' \cdot n \times \text{curl } e = \; < Se, e' > \quad (7)$$

S is thus an operator which links the values of e on Γ with the values of n × curl e on Γ. Just recall that **R** linked the values of φ on Γ with the values of n.grad Ψ on Γ.

Finally the variational formulation we obtain and that we can compare with formula (2) is the following:

$$\int_\Gamma \mu_0^{-1} \text{See}' + \int_\Omega \mu^{-1} \text{curl e curl e}' + \frac{u}{\partial t} \int_\Omega \sigma e.e' = \frac{\partial}{\partial t} \int_\Gamma e'.n \times b_0 \qquad (8)$$

3.1. DISCRETIZATION

We have seen that e, like h, is a 1-form and so we use the same tetrahedral edge elements to solve (8) and (2). It is more straightforward because we have also e as state variable on Γ and we have not to couple with a scalar potential. As everywhere in Ω the unknowns are related to the edges on Ω.

Moreover the adaptation of a code, like Trifou, using tetrahedral edge elements for h, to this new electric formulation is not very difficult because we find the same operators:

$$\int_\Omega \text{ee}' \text{ for } \int_\Omega \text{hh}' \text{ and } \int_\Omega \text{curl e curl e}' \text{ for } \int_\Omega \text{curl h . curl h}'.$$

The main difference is related to the calculation of the integral operator S, which is significantly different from the calculations of **R**, as we will see in the next paragraph.

3.2. CALCULATION OF THE INTEGRAL OPERATOR S

The starting idea is to use a single layer of a vector on Γ, whilst for **R** we used a single layer of scalar. The surfacic divergence of that vector must be equal to zero on Γ to ensure the 0 divergence of e in Ω_c.

As curl e = - $\partial b/\partial t$ we can then take:

$$e(x) = \frac{\mu_0}{4\pi} \int_\Gamma \frac{1}{|x-y|} \frac{\partial k(y)}{\partial t} dy \qquad \forall x \in \Omega_c \text{ with div } k = 0$$

The vector k can be seen as a sheet of surface current on Γ.

We have then:

$$n \times \operatorname{curl} e(x) = \frac{\mu_0}{2} \frac{\partial k(x_0)}{\partial t} + \frac{\mu_0}{4\pi} \int_\Gamma \frac{1}{|x-y|^3} n \times \left((x-y) \times \frac{\partial k(y)}{\partial t} \right) dy$$

and the boundary operator S is obtained by eliminating k from the above formulations.

To ensure that $\operatorname{div} k = 0$ on Γ, two different techniques have been used ([7] and [8]):

- we can introduce a scalar function Ψ such that $k = \operatorname{curl}(\Psi n) = \operatorname{grad} \Psi \times n$ and we solve in term of Ψ using classical nodal elements. If the domain is multiply connected, Ψ is multivalued which introduces some additional diffi-culties ([8]),

- an alternating solution is to use a normal continued interpolation function for k:

$$k = \sum_{m=1}^{3} v_m k_m \text{ with } v_m = n \times \left(\lambda i \nabla \lambda j - \lambda j \nabla \lambda i \right)$$

and k_m is the current flowing across the boundary edge ij. The problem is then that the k_m's are not independant since we have:

$$\operatorname{div} k = 0 \left(\sum_{m=1}^{3} k_m = 0 \right)$$

and we have to select the set of independant k_m's by constructing a tree of inde-pendant edges on Γ ([7]).

3.3. CONCLUSION ON THE E FORMULATION

We know that $\partial a/\partial t + e = \operatorname{grad} \varphi$ and this electric formulation in term of e is in fact the "modified vector potential method" proposed by Pillsbury in [9], or by Emson and Simkin in [10], based on the gauge $\operatorname{div}(\sigma a) = 0$.

The differences are that we avoid here the choice of a gauge, we use edge elements the advantages of which having been detailed in a previous para-graph and we use a boundary integral method outside conducting regions.

The two formulations, the electric one and the magnetic one can be seen as complementary formulation and there would be a temptation to go further and to say that in some sense they give a lower and an upper bound for some criterion to define. Unfortunately no proof of that has yet been obtained.

4. Applications concerning non-destructive testing by eddy-currents

From a mathematical point of view, non-destructive testing leads to an inverse problem: knowing the output control signal, give the parameters of the flaw. We begin working on this subject from a numerical point of view, with special attention to regularization problem, to deal with non-uniqueness and ill-conditioning of the solutions. Here we consider only the direct problems, with the method presented in the previous paragraphs. The direct problem can be very useful:

- it is possible to simulate situations where every parameter is perfectly known, which is not true in reality where the situation is badly known and complicated,

- in the reality we have access to the control signal only. With numerical modelling we can visualize what happens everywhere and at any time in the piece to control,

- we can see how a probe works before building it.

These advantages can be used in several ways:

- use the possibilities in visualization for a better understanding of the physical phenomena,

- use the possibilitiy to simulate various and perfectly known situations to establish a catalog flaws-control signals,

- use both to try to optimize existing probes or conceive new ones.

There are also some difficulties, which lead in general to a heavy consumption in CPU time. In general the presence of a flaw gives fully three dimensional problems. The movement of the probe needs to be simulated by successive calculations at different time steps and the mesh has to be very fine because of the flaw and, if it is the case, because of the differential nature of the probe.

We present two studies, one in 2D with a classical vector potential formulation and one in 3D using Trifou with the previously defined formulation.

They both concern the tubes of the steam generator in PWR nuclear reactor. The main characteristics are an enormous quantity of controls to do (more than 900.000 tubes have been inspected at this time), various flaws (wear, intergranular stress corrosion cracking...) and the use of two kinds of probe: an axial one and a rotating one.

4.1. 2D CALCULATIONS FOR AN AXIAL PROBE (C. ROSE)

The axial probe is constituted by two circular coils and moves along the axis of the tube. It often happens that flaw, roll expansion and support plate are superimposed and give superimposed signals which are very difficult to understand. In that case if the flaw is axisymmetric (transversal flaw), the situation is entirely axisymetric.

EDF is working at this time on a project to automatize the controls realized with this probe. This will be done using on expert system which needs informations about the repartition of the eddy-currents in the tube and which takes into account some hypothesis concerning the physical behaviour of the field in general.

In this context a 2D code, BIFOU, written at EDF and using a classical vector potential formulation has been widely used to check such hypothesis and to get a better understanding about what happens:

- determination of iso-phase and iso-intensity lines for the eddy-currents in a tube without flaw,

- checking of the hypothesis stating a linear relation between the intensity and phase of eddy-currents perturbed by a defect and the intensity and phase of the control signal. This can be done by cancelling a current in a finite element and checking the corresponding control signal,

- checking of the hypothesis of additivity: the global control signal due to the cancellation of the conductivity in a set of finite elements is equal to the sum of the elementary signals for each non-conducting element.

This systematic use of a 2D code for problems involving several thousands of unknowns is made possible by the optimization of it on a cray YMP. One calculation for about 5.000 unknowns corresponds to 1 sec. of CPU time.

Fig. 2 and Fig. 3 show respectively the induction lines and the control signal corresponding to the motion of the probe in front of a support plate.

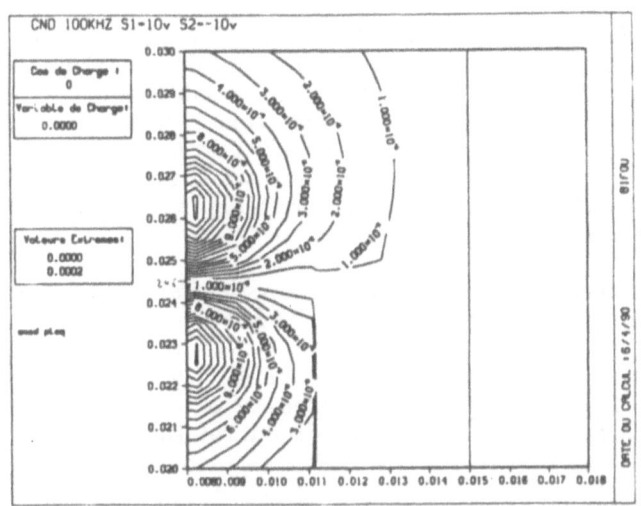

Figure 2
Induction lines.

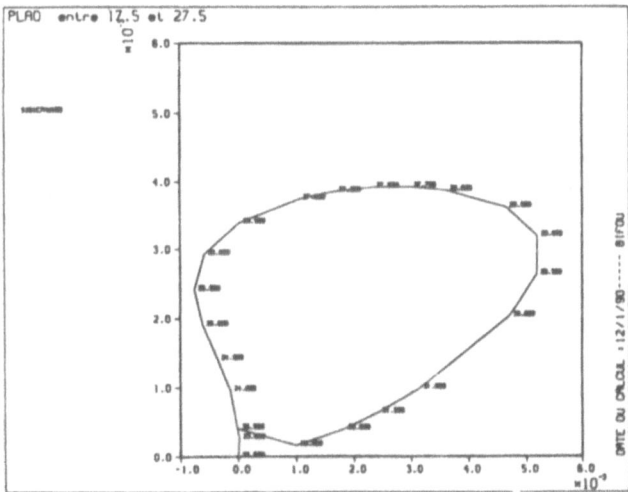

Figure 3
Control signal.

4.2. CONCEPTION OF A NEW PROBE WITH TRIFOU (J.C. VERITE) ([13])

The previous axial probe is not efficient enough to detect transversal flaws and EDF is involved in the conception of new probes.

The basic principles at the conception of an eddy-current NDT probe need a great experience. In a later stage we have to do variations around these principles to optimize it. Up to now this operation was achieved by building series of prototypes. We want to show here that this stage can be replaced by series of runs of a code and that it is faster and less expensive. To be more demonstrative we will consider only three dimensional calculations, corresponding to the particular case of the probe studied here and also to the general case when there is a flaw.

This probe has been imagined by the "Groupe des Laboratoires" in the "Service de la Production Thermique" at "Electricité de France".

It is constituted by two magnetic cores, discoid, with six poles on each core and a coil on each pole (Figure 1). These poles create a radial inducing field. The probe moves along the axis of the tube without rotating.

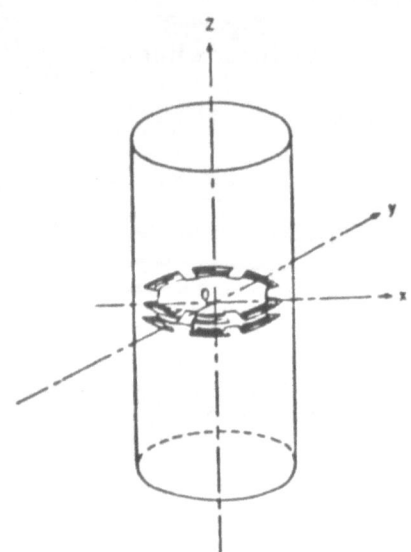

Figure 4
The probe, inside the tube, with its two six-poles
cores and a coil on each pole.

Its principles being stated, its design concerns the geometric (air-gap, number, depth and width of the poles) and electric (connection from one pole to another, frequency) parameters.

NUMERICAL RESULTS

a)   Principles

In an other work (cf. [11]) done with TRIFOU, concerning another kind of probe, our purpose was to see if it was possible to simulate control signals. Three conclusions had been made:

(i)   It is possible.
(ii) The signal is created by the movement of the probe. To simulate this movement we have to do a certain number of calculations ($\sim$ 10-20) corresponding to different positions of that probe.
(iii) When the signal is differential the mesh must be very regular to avoid a differential effect due to variations of it.

As a consequence it is very expensive to obtain a control signal with a 3D numerical calculation. We propose the three following stages combining efficiency and reasonable CPU times:

1)   Parametric studies, involving a large number of program runs without flaw and without movement. This stage gives a precise knowledge of fields and currents induced by the probe in the piece to control. It is possible to visualize their repartition by postprocessing and to do a first selection of parameter values.

2)   Studies with flaw but without movement of the solutions selected in the previous stage. At this point for the few acceptable solutions we have to see more precisely how the current lines are modified by the presence of a flaw. Again visualization by postprocessing is essential.

3)   For the one or two best solutions, study with flaw and movement. The problem here is to simulate the control signal to see if the previous conclusions on the current distributions, obtained by local considerations, are confirmed by the output signal.

b)   Results

Only the first stage has been realized. Taking symmetries into account, the domain consists of one-half of a pole and one piece of tube, on 50 mm along the axis . For this bounded domain each run has been done using about 4000 finite elements, corresponding to 4 minutes of CPU time on Cray 1S.

The influence of the following parameters has been studied: the frequency, the supply of two successive coils on a core, of two corresponding coils on each core, the distance between the two cores, the width of a core, the distance between two successive poles. Choice criteria consist in shape and area of current distribution.

On figure 5 and 6 we can see in perspective the frame of the meshed domain which consist in one-half of a pole and the corresponding part of the tube.

We can see also the current density at the center of each finite element, near the inner face of the tube on figure 5 and a quarter of a period later on the outer face on figure 6. Figure 7 shows the flux of current density through three cutting planes. This gives an idea of the "detectability" of a flaw in these planes.

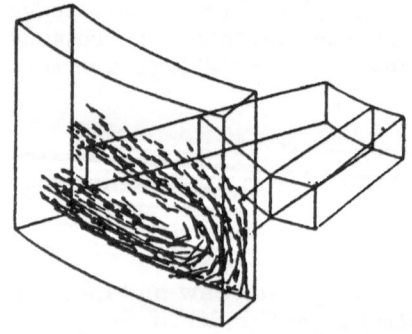 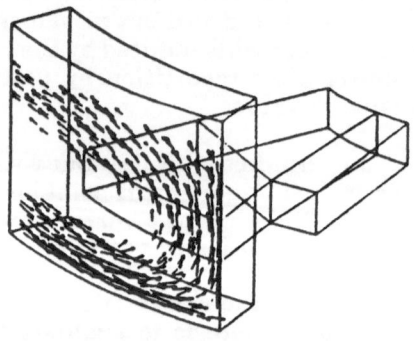

Figure 5
Currents near the inner
face of the tube Ψ = 90°.

Figure 6
Currents near the outer
face of the tube Ψ = 90°.

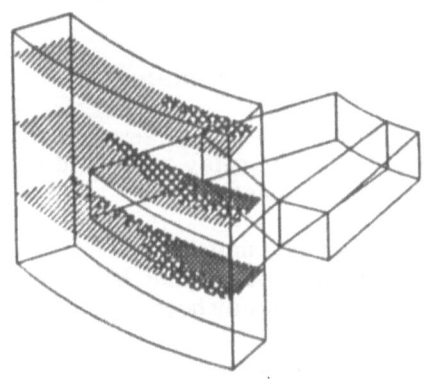

Figure 7
Flux of current density through cutting planes.

5.   An application concerning metallurgy (P. Chaussecourte)

In metallurgy the rolling or laminating process is the transformation of thick
products ($\simeq 200$ mm) obtained by melting into thin sheets of steel (from 2 to 15
mm). This is done using a set of operations during which the metallurgical
product, on a train, passes first inside a set of reheating furnaces giving a
proper temperature ($\simeq 1200$-$1300°$), then through roughing mills giving a
product of about 30 to 50 mm thick, then on a delay table and finally through
finishing mills giving the end product with the desired thickness.

The temperature decreases continuously along this process and becomes less
and less homogeneous. In particular the edges of the product become cooler
and cooler compared with the mean temperature of the block.

Several problems arise:

     - an overheating of the whole block is necessary to get a proper tempe-
rature of the edges,

154

    - formation of metallurgical heterogeneities giving important losses for the final product,

    - a wear of the cylinders for laminating.

To overcome these difficulties it seems necessary to realize a localized heating of the edges of the product on the delay table. Induction seems to be particularly appropriate to realize this heating.

Figure 8 shows a picture of an induction device conceived in collaboration between EDF, ALSTHOM-ROTELEC and IRSID. At this time it has been marketed and some ones have already been sold all over the world.

It is composed of a C-shape magnetic circuit with the metallurgical product passing through the air-gap and of two inducing coils at the extremities of the magnetic circuit.

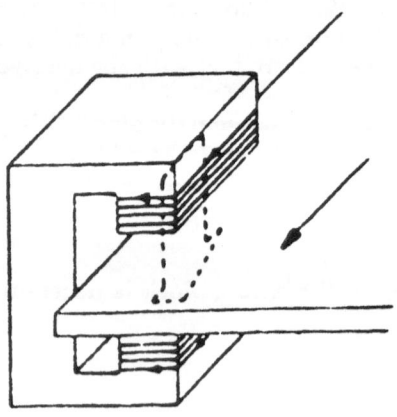

Figure 8
The induction device.

Numerical modelization presents several advantages at the conception stage:

    - to calculate the distribution of the heat in the block function of the positionning of the C-shape circuit around the edge,

    - to determine the active power transmitted to the block function of the air-gap between the poles and of the positionning of the magnetic circuit,

    - to evaluate the magnetic induction in the magnetic circuit, in particular to avoid, as far as possible, the magnetic saturation which gives extra losses.

So the modelization should have to optimize the device at the conception stage. In fact the use of a numerical software, TRIFOU, happened at a later stage, when a prototype was already realized. Its aims were then to check that the choices that had been choosen were adequate and to give a better knowledge of how it works.

Figure 9 shows the hexahedral mesh from which we get a tetrahedral mesh by dividing each hexahedron in 24 tetrahedron. Using the symmetries we used 9500 tetrahedra for one fourth of the device including the magnetic circuit and a sufficient part of the block to heat. This mesh involves 10000 unknows and the problem need 15 mn of CPU-time on a Cray XMP.

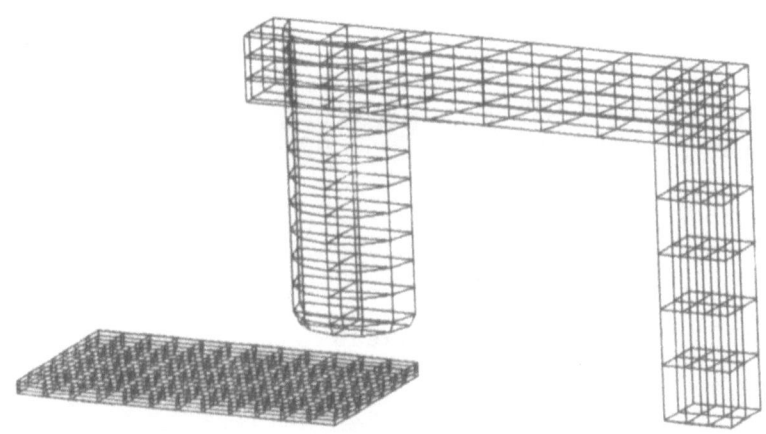

Figure 9
Hexahedral mesh.

We can see on figure 10 the current density induced in the load. We can then calculate Joule losses, necessary to calculate the temperatures. The numerical calculation gives also the total active power transmitted to the load, giving the efficiency of the process.

We can also calculate and display the magnetic induction everywhere in the magnetic circuit, showing where and for what inducing current saturation can occur.

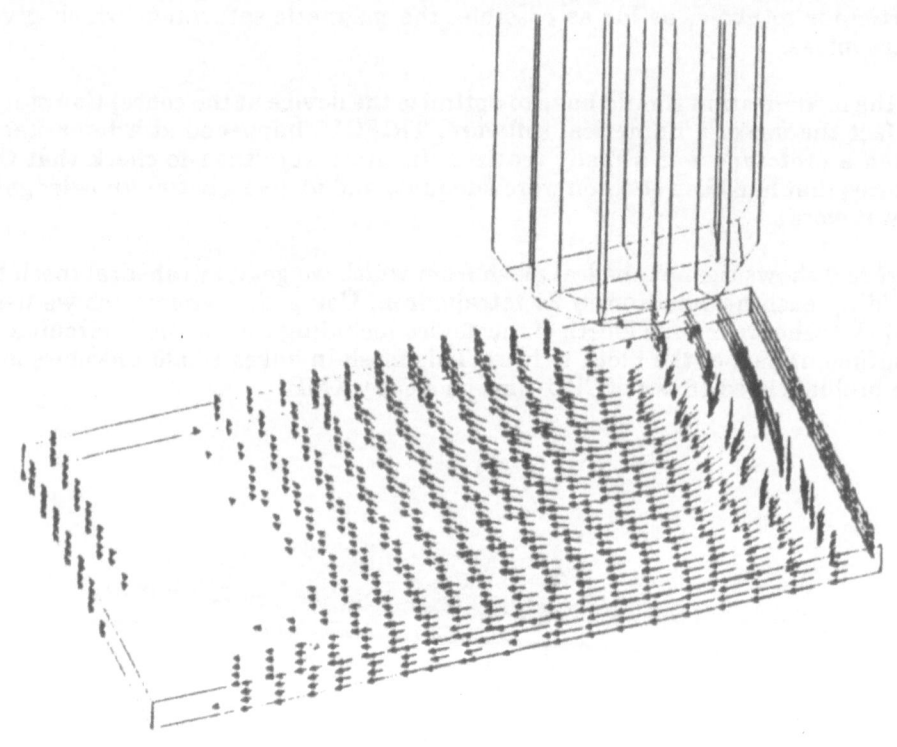

Figure 10
Current density in the load.

The numerical simulation gives thus answers to very practical questions asked to the engineer who conceives a device. It is important to note that if the computer costs can seem rather important for such a study, they are negligible compared with the cost for building a prototype. We can say that in such a problem numerical simulation is fully operational in the conception stage.

6. An application about currents induced in metallic structures of Tokamaks (P. Chaussecourte) ([12])

We briefly mention here an important work which has been done in collaboration between EDF and Ispra Centre and which will be more detailed in an other part of this course.

The international fusion project, ITER, involves the design of a large experimental Tokamak machine, where a toroidal current carrying plasma is driven and confined. Accidental events that must be considered during the long pulse controlled burn of the plasma ring are the plasma disruptive instabilities. They are related to failure conditions of the machine equipment for the plasma feedback stabilization.

During such scenarios, eddy currents are induced in all the metallic components exposed to the transient magnetic fields. The knowledge of these currents is essential for the estimate of the associated electromagnetic forces and stresses.

We have used the TRIFOU code, involving the formulation detailed at the beginning of this work, to determine the eddy-current distribution in the first wall of the ITER Tokamak during a plasma disruption (Figure 11).

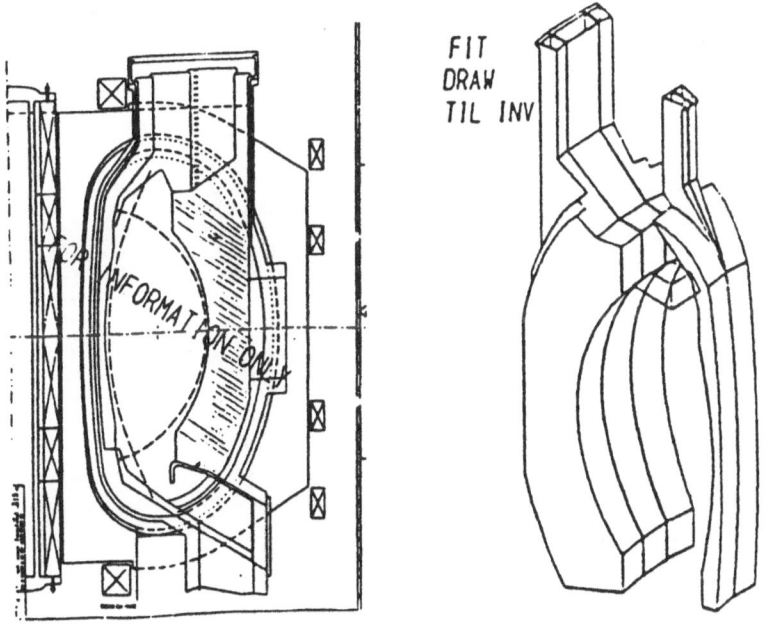

Figure 11

ITER REACTOR - FIRST WALL OUTBOARD SEGMENTS

There were several specific difficulties in this application where the conducting domáin to mesh is a set of more than 30 hollow boxes representing the first wall. The air gap between boxes is very small and the walls have different thicknesses, some ones very small giving constant currents in the thickness and some others thick enough to greatly justify such a 3D calculation. Using the symmetries we mesh only one quarter of a box.

The main difficulties are the small air gaps and thin walls needing a special care for the calculation of integrals in the determination of the outer stiffness matrix.

This is also one of the first application involving transients instead of steady-state time variations.

A quarter of a box was meshed using 18000 tetrahedra giving 20000 unknowns (corresponding to 2,5 millions of elements for the whole structure). A plasma disruption of 20 ms has been simulated by 200 time steps covering 40 ms, at the end of which every induced current has vanished. The total CPU time for calculating eddy-currents and forces was more than 3 hours on a CRAY-XMP (Figures 12 and 13).

Figure 12 shows the induced current in a symmetry plane and the figure 13 the Laplace body forces, function of time, inside the different parts of the box.

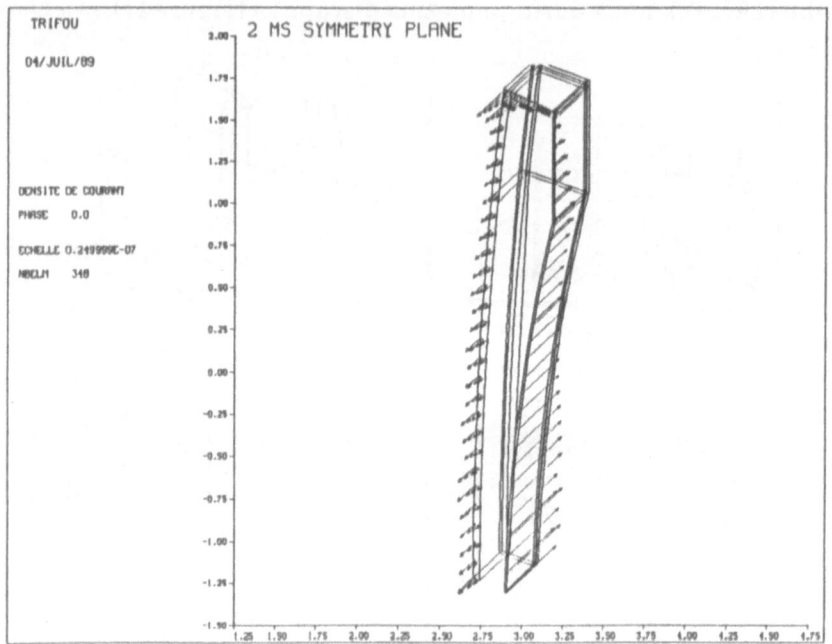

Figure 12
Induced current in a symmetry plane.

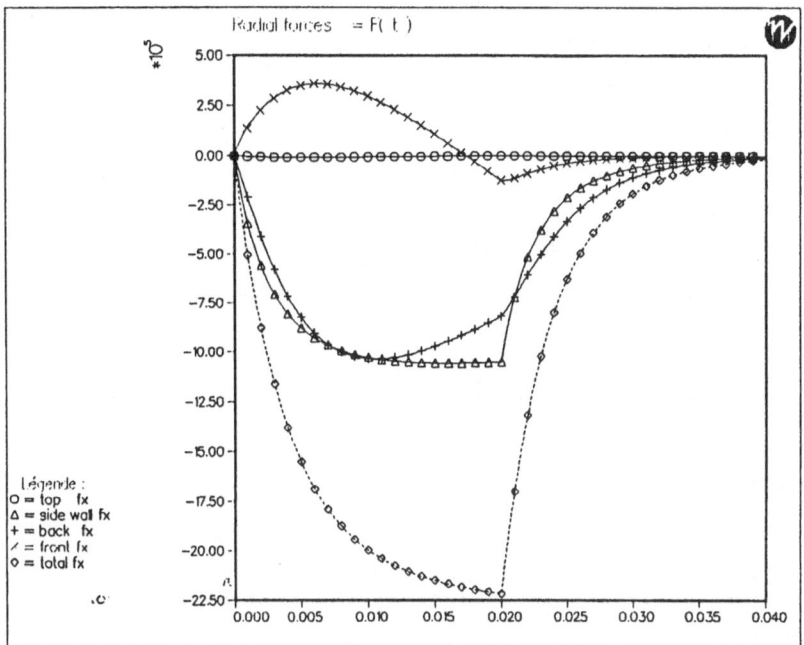

Figure 13
Laplace body forces.

7.   Conclusion

The recent improvements in mathematical formulation, software enginee-ring and power of computers allows us to do some general conclusions, enligh-tened by the three previous applications.

We have reached a level of development for the 3D eddy-current codes allowing to solve a very large panel of problems with very different geometries and physical properties.

We are now able to treat real industrial problems needing 3D calculations with a good probability to save money compared with building of series of prototypes. To do that numerical simulation has to be utilized at the conception stage.

It is possible to model a whole simple device (NDT probe, induction heating apparatus) or to do a local study on a part of a complicated device (Tokamaks).

160

BIBLIOGRAPHY

[1]   A. BOSSAVIT, J.C. VERITE.
        A mixed FEM-BIEM method to solve 3D eddy-current problems.
        IEEE MAG 18 (2), 431 (1982).

[2]   A. BOSSAVIT.
        Introduction théorique à TRIFOU.
        E.D.F. - Internal Report HI-4759/00 (1985).

[3]   A. BOSSAVIT.
        Coupling finite elements and boundary elements in eddy-current
        computations.
        IUTAM Symposium on Advanced Boundary Element Methods.
        San-Antonio - Texas (April 1987).

[4]   A. BOSSAVIT.
        Mixed finite elements and the complex of Whitney forms.
        The Mathematics of finite elements and Applications VI
        (J.R. Whiteman ed.).
        Acad. Press (London), (1988).

[5]   H. WHITNEY.
        Geometric integration theory.
        Princeton U.P. (1957).

[6]   Z. REN, F. BOUILLAULT, A. RAZEK, J. C. VERITE.
        Comparison of different boundary integral formulations when cou-
        pled with finite elements in three dimensions.
        IEEE Proc., Vol. 135, A, n° 8, pp. 501-507.

[7]   Z. REN, F. BOUILLAULT, A. RAZEK, A. BOSSAVIT, J.C. VERITE.
        A new hybrid model using electric field formulation for 3D eddy-
        current problems.

[8]   J.C. VERITE.
        Computation of eddy-currents on the alternator output conductors by
        a finite element method.
        Electrical Power and Energy systems, 1, 3, pp. 193-198, (1979).

[9]   R.D. PILLSBURY.
        A three dimensional eddy-current formulation using two potentials:
        the magnetic vector potential and total magnetic scalar potential.
        IEE Trans., MAG-19, (6), pp. 2284-2287, (1983).

[10]   C.R.I. EMSON, J. SIMKIN.
        An optimal method for 3D eddy-currents.
        IEEE Trans, MAG-19, (6), pp. 2450-2452, (1983).

[11]   J.C. VERITE.
        A coil over a crack (Results for benchmark Problem 8 of TEAM
        workshop).
        To appear in COMPEL - Spring 1990.

[12]   P. CHAUSSECOURTE, A. BOSSAVIT, J.C. VERITE,
        Y.R. CRUTZEN.
        3D computer simulations of the electromagnetic transients induced in
        the Tokamak first wall by a plasma disruption.
        COMPUMAG Conference, Tokyo, (September 1989).

[13]   M. CARRE, J.C. VERITE.
        Use of a 3D eddy-current code for optimizing an NDT probe.
        Trans. on SMIRT-9, Lausanne, (August 1987).

COMPUTATIONAL METHODS FOR THE ELECTROMAGNETIC ANALYSIS IN FUSION DEVICES

R. ALBANESE (1) and G. RUBINACCI (2)

1) Istituto di Ingegneria Elettronica, Università di Salerno,
   I-84081 Baronissi (Salerno), Italy
2) Dipartimento di Ingegneria Elettrica, Università di Napoli
   "Federico II", Via Claudio 21, I-80125 Napoli, Italy

ABSTRACT. The study of electromagnetic phenomena in a Tokamak device is essential for the comprehension of its physical behavior and for the design of future reactors.

From the electromagnetic point of view, three interacting systems can be identified in a Tokamak: coils, conducting structures and plasma. A full description of the problem in a three-dimensional geometry presents formidable numerical difficulties due to the complexity of the interacting phenomena from both geometrical and physical points of view. On the other hand, it is often sufficient to study in detail a single aspect of the problem, depending on the objectives of the analysis (electromechanical design, simulation of the experiment, design of the control system).

When the plasma is the objective of the analysis, circuits and passive structures are modeled in a simplified way, while in the case of the three-dimensional eddy current computation, the plasma plays the role of a source term. Of course, to get satisfactory results, it is often necessary to solve both models repeatedly.

In this work we will review the basic model which describes the main features of the plasma subsystem (in the framework of the ideal MHD theory), and the subsystem including the massive structures. In this case, an integral formulation of the problem will be reviewed with particular reference to its relevant numerical aspects.

1. INTRODUCTION

Controlled thermonuclear fusion promises to provide an inexhaustible source of energy with a very limited environmental impact. For this reason, it represents one of the principal objectives of the present scientific research.

To achieve nuclear fusion, it is necessary to confine a fully ionized mixture of deuterium and tritium (plasma) having particle density n, at a certain temperature T for a certain time τ_E. To have the nuclear fusion self-sustained, it is necessary to achieve T > 10keV

161

Y. R. Crutzen et al. (eds.), Industrial Application of Electromagnetic Computer Codes, 161–187.
© 1990 *ECSC, EEC, EAEC, Brussels and Luxembourg.*

and n τ_E T $> 5 \times 10^{21} m^{-3}$ s keV.

Two principal confinement methods are now under investigation. The inertial approach intends to confine a high density plasma ($n \cong 10^{30} m^{-3}$) for a very short time($\tau_E \cong 10^{-10}$ s). The magnetic confinement utilizes a low density plasma (n $\cong 10^{20} m^{-3}$); of course, in this case, the required confinement time is much longer ($\tau_E \cong$ 1s).

Among the various possible choices, the approach that achieved the best results is the toroidal magnetic configuration in Tokamaks [1].

From the electromagnetic point of view, three interacting systems can be identified in a Tokamak device: coils, conducting structures and plasma. A schematic view of a Tokamak is illustrated in Fig. 1.

Three sets of coils (most of them superconducting) can be identified:
- Ohmic heating (OH) coils, which provide the electric field for plasma current inductive drive;
- Toroidal Field (TF) coils, which create the toroidal field necessary for confinement and Magneto-Hydro-Dynamic (MHD) stability
- Poloidal Field (PF) coils, which provide the poloidal field for the open and closed loop control of plasma position and shape.

The main conducting structures surrounding the plasma are:
- the plasma facing components (First Wall and Blanket modules), which also play the role of heat exchangers;
- the vacuum vessel which also play a shielding action;
- the casings of the superconducting coils.

The electromagnetic behavior of a Tokamak plasma is that of a basically axisymmetric deformable conductor, whose evolution is determined by the electromagnetic force density J×B and the transport phenomena. The study of the electromagnetic phenomena in a Tokamak device is essential for the comprehension of its physical behavior and for the design of future reactors.

Starting from a desired plasma configuration at flat top, the first step in the project of a Tokamak concerns the design of the TF, OH and PF equilibrium coils. The main problem in this context is to find a set of currents producing a prescribed magnetic field configuration. For this purpose, typical approaches to solve inverse problems are used. Of course, iterations are necessary if geometrical, mechanical, thermal, or technological constraints are not satisfied.

The basic electromagnetic design includes some more steps involving the assessment of the electric parameters of the conducting structures (e.g. the toroidal resistance of the vacuum vessel) and the characterization of the control system (e.g. the choice of the power amplifiers).

However, the most important task of the electromagnetic studies in a Tokamak is related to the analysis and verification of the behavior of its components during the fast transitions determined by current and magnetic field changes. Electromagnetic transients occur during normal operation (e.g. plasma start-up, shape and position control, shut-down) and accidental conditions (e.g. plasma disruptions, failure

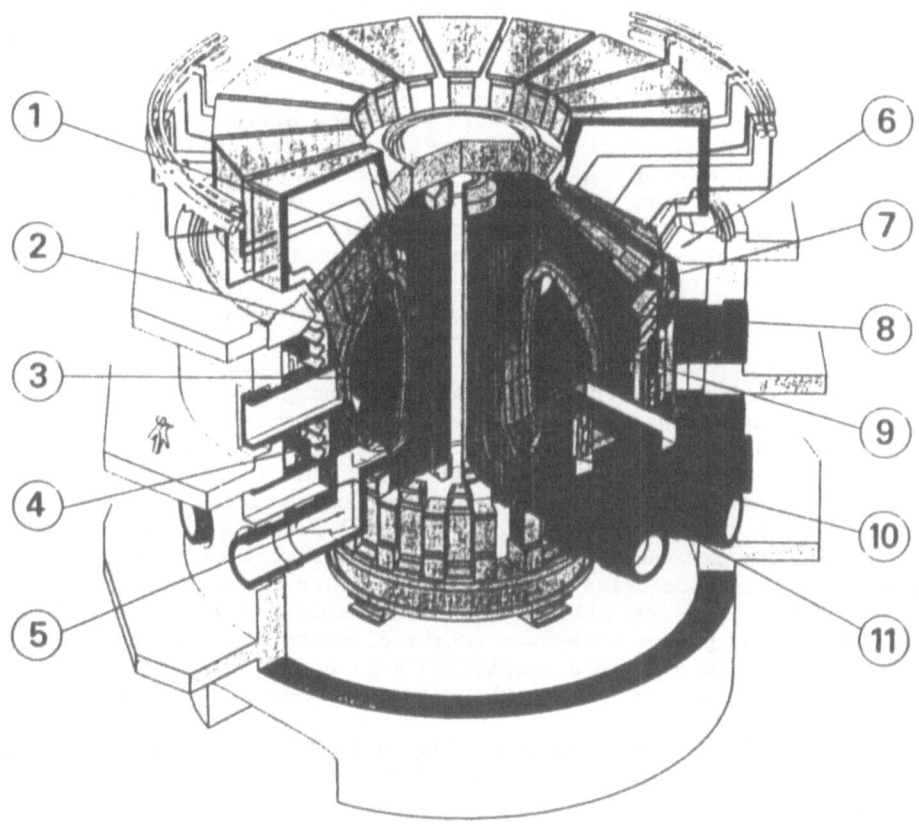

1- INNER PF. COILS
2- BLANKET
3- PLASMA
4- VACUUM VESSEL-SHIELD

5- PLASMA EXHAUST
6- BIOLOGICAL SHIELD-CRYOSTAT
7- ACTIVE CONTROL COILS
8- TOROIDAL FIELD COILS

9- FIRST WALL
10- DIVERTOR PLATES
11- OUTER PF. COILS

Figure 1. The NET (Next European Torus) Tokamak.

of a superconducting coil). The main effects produced by these
transient phases are induced currents and voltages in structures and
coils which can give rise to field penetration delays, forces,
insulation damages, arcing and heat.

These transient electromagnetic phenomena are so complex from both
geometrical and physical points of view, that it is practically
impossible to develop a single methodology which results to be
satisfactory for each particular aspect. A full description of the
problem should take into account the simultaneous evolution of three
different but interacting phenomena in three different subsystems:
circuits (including TF, OH and PF coils) fed by open or closed loop
amplifiers; eddy currents in the massive conducting structures; the
electromagnetic behavior of the plasma.

Therefore, simplified models and related computational methods
were proposed to study the transient electromagnetic phenomena, paying
particular attention to a single subsystem (circuits, massive
conducting structures, or plasma). For example, a circuit code can
provide an approximate but agile tool for the design of the control
system; in this case the plasma and the massive conducting structures
are schematized, in a crude way, as a set of suitable lumped parameters
entering into the dynamic system. These parameters can be computed by
means of the methodologies focused on the other two systems.

In this work we will analyze the models describing the main
features of the plasma subsystem (where circuits and passive structures
are taken into account in a simplified way) and the subsystem including
the massive structures (where the plasma current plays the role of a
source term).

These methodologies substantially differ from each other not only
for their specific focus of interest (circuits, eddy currents, plasma),
but also for their primary objectives (calculation of electromechanical
loads, design of the control system, simulation of the experiment in
its global aspects).

2. MATHEMATICAL MODEL OF THE PLASMA

2.1. The MHD model

The core of a nuclear fusion device is represented by the plasma. The
study of the macroscopic behavior of fusion plasmas is a considerably
complex problem, presenting some aspects which are still now not
completely understood. A good description of the macroscopic behavior
is provided by the ideal MHD model [2]:

$$\frac{\partial \rho}{\partial t} + \nabla \cdot \rho \mathbf{v} = 0 \tag{1}$$

$$\rho \frac{d \mathbf{v}}{d t} = J \times B - \nabla p \tag{2}$$

$$p = p(\rho) \tag{3}$$

$$E + v \times B = 0 \tag{4}$$

coupled to the quasistationary Maxwell equations:

$$\nabla \times E = -\frac{\partial B}{\partial t} \tag{5}$$

$$\nabla \times H = J \tag{6}$$

$$\nabla \cdot B = 0 \tag{7}$$

the constitutive equations

$$B = \mu H \tag{8}$$

$$J / \sigma = E + E_{ext} \tag{9}$$

with suitable boundary and initial conditions.
In these equations, E is the electric field, E_{ext} the externally applied electric field, B the magnetic flux density, H the magnetic field, J the current density, μ the magnetic permeability, σ $(=1/\eta)$ the electric conductivity, ρ the mass density, v the fluid velocity and p the pressure. Eq. (9) has been written with reference to the conductors outside the plasma.

Despite its elegant simplicity, the MHD model is still too difficult to be solved in most of the geometries of interest.

An analysis of the problems involved in the study of the MHD model is outside the scope of this work. Here we just recall the main features of the problems which have a direct interaction with the engineering design of a Tokamak device: the plasma equilibrium and the plasma evolution on a time scale long compared to the characteristic MHD time.

2.2. Plasma equilibrium

The first class of engineering problems is related to the design of the magnetic configurations which assure stable plasma equilibria. The basic equations to be solved in this case are the time-independent form of the full MHD equations with $v = 0$:

$$J \times B = \nabla p \tag{10}$$

$$\nabla \times H = J \tag{11}$$

$$\nabla \cdot B = 0 \tag{12}$$

$$B = \mu H \tag{13}$$

In the cases of interest, using a cylindrical coordinate system (R, φ, Z), the hypothesis of axisymmetry allows us to write:

$$B = \frac{\nabla\psi}{R} \times e_\varphi + \frac{f}{R} e_\varphi \tag{14}$$

$$J = \frac{\nabla(f/\mu)}{R} \times e_\varphi + L\psi \, e_\varphi \tag{15}$$

where e_φ is the unit vector along the toroidal direction,

$\psi = \int_0^R R' \, B_z(R', Z) \, dR'$ is the poloidal flux per radian,

$f/\mu = \int_0^R R' \, J_z(R', Z) \, dR'$ is the flux per radian of the poloidal current density, and

$$L\psi = - \frac{\partial}{\partial R} \left(\frac{1}{\mu R} \frac{\partial\psi}{\partial R} \right) - \frac{\partial}{\partial Z} \left(\frac{1}{\mu R} \frac{\partial\psi}{\partial Z} \right) \tag{16}$$

Expressing B and J by means of eqs. (14)-(15), eq. (10), i.e. the momentum equation inside the plasma, where $\mu = \mu_0$, can be rewritten as:

$$\nabla p = \frac{L\psi}{R} \nabla\psi - \frac{f \nabla f}{\mu_0 R^2} \tag{17}$$

The momentum equation (10) also implies

$$\nabla p \cdot J = 0, \qquad \nabla p \cdot B = 0 \tag{18}$$

Hence, the current density and magnetic field lines lie on the surfaces described by p=constant, which are simultaneously magnetic and current surfaces. The hypothesis of axisymmetry implies that ψ and f are constant on a magnetic surface. Then, if the function ψ is chosen as a variable describing the family of nested magnetic surfaces, it follows that:

$$\nabla p = \frac{dp}{d\psi} \nabla\psi, \qquad \nabla f = \frac{df}{d\psi} \nabla\psi \tag{19}$$

Inserting the expressions (19) into the equilibrium equation (17), one obtains the Grad-Shafranov equation [3, 4]:

$$L\psi = J_\varphi = R \frac{dp}{d\psi} + \frac{f}{\mu_0 R} \frac{df}{d\psi} \tag{20}$$

where $p(\psi)$ and $f(\psi)$ are two functions that can be prescribed on the basis of experimental observations (indeed they are determined by the prehystory of the discharge). In Fig. 2 a typical plasma equilibrium configuration is shown.

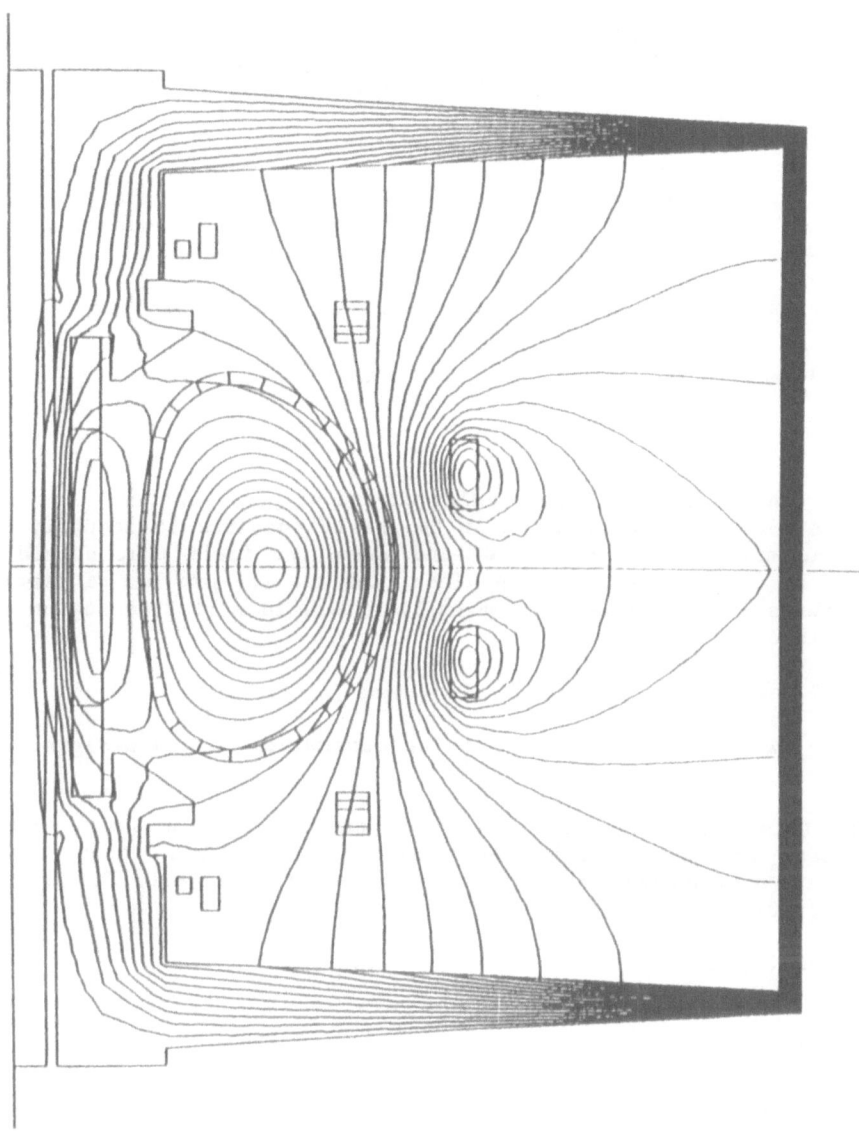

Figure 2. A JET (Joint European Torus) equilibrium configuration.

It can be shown that the plasma cannot be magnetically confined solely by its own current; an external coil system is necessary. From the point of view of the electromagnetic design, two problems related to the study of the plasma equilibria can be identified:

Problem I

The system of external PF coils is to be determined, for a desired plasma configuration, i.e. the plasma cross section S_P and the plasma current density profile $J\varphi$ (ψ, R) are supposed to be known.

In this case, one has to solve the fixed boundary problem:

$$L\psi - J_\varphi - R \frac{dp}{d\psi} + \frac{f}{\mu_0 R} \frac{df}{d\psi} \qquad \text{inside } S_P \qquad (21)$$

$$\psi - \psi_0 \qquad \text{on } \partial S_P \qquad (22)$$

which is linear only for a particular choice of $p(\psi)$ and $f(\psi)$.

Once ψ has been determined in S_P, it is possible to identify the plasma current contribution to ψ, and hence the field that should be supplied by the external sources. Then, the inverse problem of finding the currents giving an assigned magnetic field in a limited region can be formulated [5].

Problem II

The plasma configuration is to be determined for an assigned set of external coils

In this case, the plasma current density profile $J\varphi$ is still prescribed, but the position and shape of the plasma boundary is unknown and must be determined through the solution itself. The problem can be formulated as follows:

$$L\psi - J\varphi (\psi, R) \qquad \text{for } \psi \geq \psi_b \qquad (23)$$

$$L\psi - J\varphi_{ext}(R, Z) \qquad \text{for } \psi < \psi_b \qquad (24)$$

$$\psi_b - \sup_{\partial S_0} \psi \qquad (25)$$

being ψ_b the value of ψ at the plasma boundary, defined as the outermost flux line inside the limiter ∂S_0.

2.3. Plasma evolution

Neglecting rapid inertial effects (i.e. for a time scale long compared to the characteristic MHD time $\tau \approx 1\mu s$), the plasma evolution can be simulated as a series of quasistatic equilibria still described by the Grad-Shafranov equation (20). In this case however the plasma current density profile, related to the profiles of p and f, is a function of time and additional equations are needed to close the system [6, 7, 8, 9]. Several possible models can be used. As a first choice, the time evolution of the current profile can be prescribed. A further improvement toward an MHD consistent description of the plasma evolution is obtained by determining $f(\psi, t)$ on the basis of the knowledge of the profile of the plasma resistivity parallel to the magnetic field. In this approach the pressure profile $p(\psi, t)$ is prescribed, thus uncoupling the model from the particle and energy transport equations.

In the ideal MHD model the plasma is assumed to be a perfect conductor and hence during the evolution both the toroidal and poloidal magnetic fluxes are conserved. This condition determines $f(\psi, t)$. In fact, in this zero resistivity limit, it can be shown that the quantity $d\Phi/d\psi = -2\pi\ q(\psi)$ is constant in time, having defined $\Phi(\psi)$ as the flux of the toroidal magnetic field trough the surface ψ=constant. Moreover $f(\psi)$ is related to $q(\psi)$ in the following way:

$$f(\psi) = 2\pi\ q(\psi)/ \oint dl_P/(R\ |\nabla\psi|\ ) \tag{26}$$

where the circulation of $1/(R\ |\nabla\psi|)$ is performed along the poloidal cross section of a surface ψ=constant.

Substituting this expression for f in the r.h.s. of eq. (20), one obtains an integro-differential equation which describes the flux conserving plasma evolution [10].

The currents in the massive axisymmetric conductors are determined by means of Faraday's law:

$$J\varphi = -\frac{\sigma}{R}\frac{\partial\psi}{\partial t} \tag{27}$$

The currents in the PF coils can be taken into account by the loop current method. Let $J_k\ (R, \varphi, Z)$ the current density associated to the k-th loop current, where the φ-dependence stresses the possibility of having non-axisymmetric electric connections. Thus, the toroidal current density $J\varphi$ in the PF coil region is given by:

$$J\varphi = \sum_{k=1}^{N}\ I_k(t)\ q_k(R, Z) \tag{28}$$

where N is the number of independent loops and q_k are the average values of $e_\varphi \cdot J_k$ in $0 \leq \varphi \leq 2\pi$.

In order to determine the behavior of the new unknowns $I_k(t)$, other equations should be introduced, relating the $I_k(t)$ to the poloidal flux equation in the PF coil region.

These equations [11] can be obtained by Ohm's law:

$$J / \sigma = E + E_{ext} \tag{29}$$

Multiplying both sides by J_i and integrating over the PF coil region Ω_{PF}, it results:

$$- \int_{\Omega_{PF}} J_i \cdot E \ dV + \sum R_{ik} \ I_k(t) = V_i \tag{30}$$

where $V_i = \int_{\Omega_{PF}} J_i \cdot E_{ext} \ dV$ is the externally applied e.m.f. to the i-th loop and $R_{ik} = \int_{\Omega_{PF}} J_i \cdot J_k / \sigma \ dV$ is the mutual resistance between i-th and k-th loops. The inductive contribution $- \int_{\Omega_{PF}} J_i \cdot E \ dV$ can be written as:

$$- \int_{\Omega_{PF}} J_i \cdot E \ dV = 2\pi \int_{\Sigma_{PF}} q_i \ \frac{\partial \psi}{\partial t} \ dS + \sum M_{ik}^{ext} \ \frac{dI_k}{dt} \tag{31}$$

The first term on the r.h.s. takes into account the axisymmetric part of the windings. The second term includes the external connections; in fact, M_{ik}^{ext} represents the contribution to the mutual inductance between i-th and k-th loops, due to their non axisymmetric branches.

The presence of non axisymmetric massive structures can be taken into account by a proper circuit schematization similar to the one associated to the PF coils. The relevant equivalent parameters can be obtained with the help of fully 3-D eddy current codes and model reduction procedures [12].

In summary, the time evolution of the plasma current distribution can be modeled as a series of quasistatic 2-D axisymmetric MHD equilibria, evolving in the presence of active and passive external conductors. The most important mechanism for this evolution is given by the diffusion of the magnetic field, leading to a typical eddy current problem coupled with the MHD plasma equilibrium. In conclusion, the plasma evolution, in the limit of zero plasma resistivity, can be described by the following set of axisymmetric equations:

$$L\psi = J\varphi \tag{32}$$

$$2\pi \int_{\Sigma_{PF}} q_i \ \frac{\partial \psi}{\partial t} \ dS + \sum M_{ik}^{ext} \ \frac{dI_k}{dt} + \sum R_{ik} \ I_k(t) = V_i \ , \quad i=1, N \tag{33}$$

where:

$$J\varphi = - \frac{\sigma}{R} \frac{\partial \psi}{\partial t} \qquad \text{in the massive axisymmetric conductors} \tag{34}$$

$$J\varphi = \sum_{k=1}^{N} I_k(t) \, q_k(R, Z) \qquad \text{in the PF coil region} \qquad (35)$$

$$J_\varphi = R \frac{\partial p}{\partial \psi} + \frac{f}{\mu_o R} \frac{\partial f}{\partial \psi} \qquad \text{inside the plasma region } S_P \qquad (36)$$

defined as the region where:

$$\psi \geq \psi_b, \quad \psi_b = \sup_{\partial S_o} \psi$$

with $f(\psi, t) = 2\pi \, q(\psi) / \oint \, dl_P / (R \, |\nabla\psi|)$

and given $p = p_o \, (\psi, t)$ and $q = q_o(\psi)$.

This set of non linear integro-differential equations is solved in the code PROTEUS [6, 9] by means of the Finite Element Method, using a Galerkin procedure in order to obtain a non linear set of ordinary differential equations to be advanced in time. The constraint on the flux conservation inside the plasma is imposed using an integral weak formulation (in the form of a compatibility constraint over the plasma volume), in which the cumbersome computation of integrals along the field lines is avoided. An MHD simulation of a plasma evolution computed via the PROTEUS code is shown in Fig. 3.

The main limitation of this model of the plasma evolution is its restriction to axisymmetric geometries. This constraint is due to the extreme complexity of a fully 3-D model and consequent limits of memory occupation and CPU time. On the other hand, the hypothesis of axisymmetry is acceptable for the modeling of the plasma in a context related to the engineering simulation of the whole experiment in its global aspects, covering the entire range of start-up, disruptions, position and shape control.

3. MODEL AND SCHEMATIZATION OF THE THREE-DIMENSIONAL CONDUCTING STRUCTURES

The transient electromagnetic fields in a Tokamak device give rise to eddy currents in the metallic structures of the machine. These eddy currents can produce very important effects on the plasma dynamics, on the superconducting coils and on the structures themselves.

Eddy currents arise during both normal operation and accidental conditions. The main effects produced during normal operation are the delay in the penetration of the electromagnetic fields inside the vacuum vessel, losses in the superconductors, and influence - not always negative - on the control of plasma position and shape. The principal effects produced during accidental conditions like plasma disruptions are the electromagnetic loads on the structures (due to the interaction of the eddy currents with the strong magnetic fields existing in a Tokamak) and the shielding action toward the external PF and TF coils.

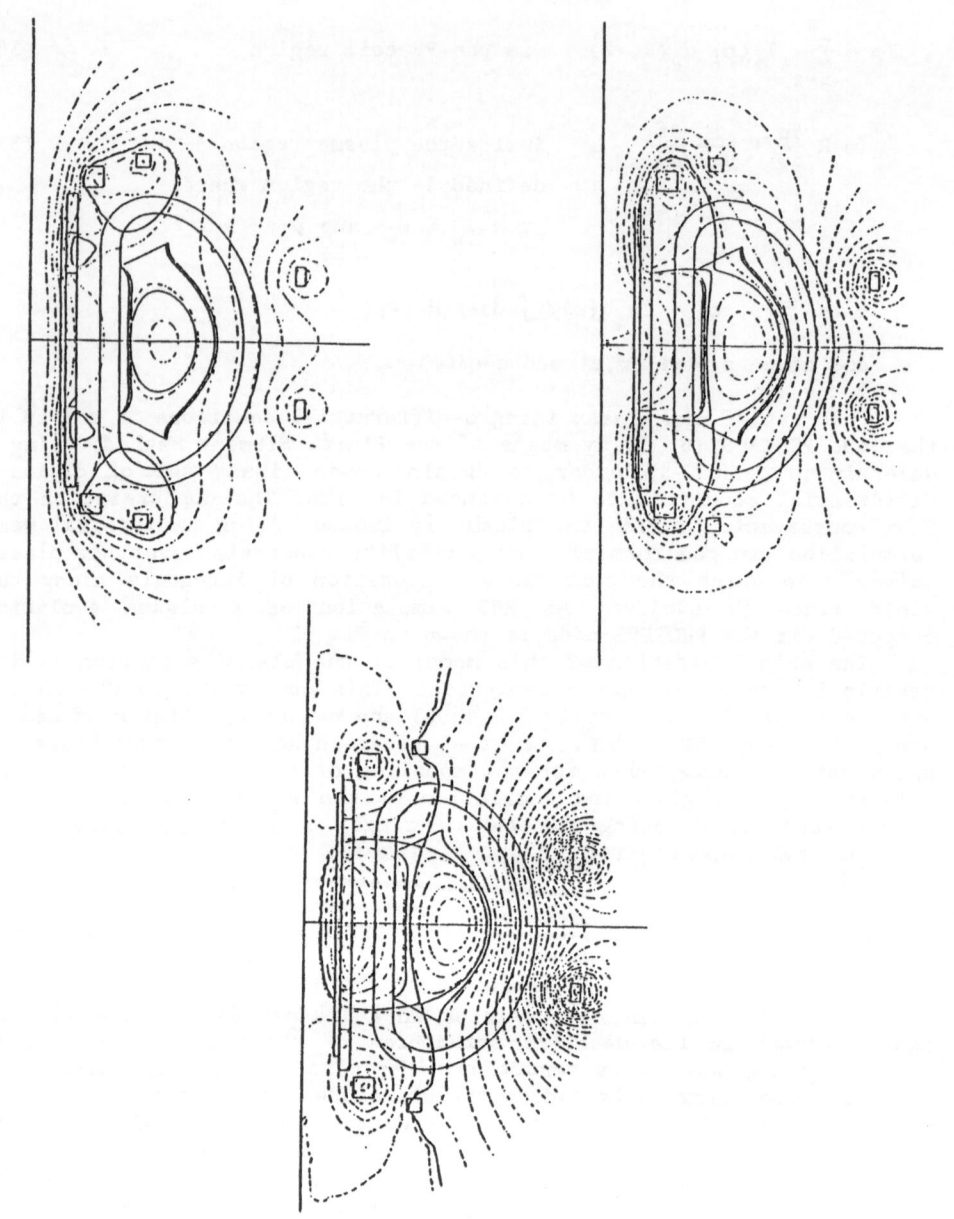

Figure 3. Simulation of plasma evolution in NET.

The choice of model and schematization of the conducting structures for the electromagnetic analysis in a fusion reactor is mainly determined by the objectives of the study.

For instance, axisymmetric or even more crude models of the structure are also adequate for the study of basically axisymmetric phenomena like the penetration of the electromagnetic fields inside the vacuum chamber or the control of the plasma shape.

However, more detailed and fully 3-D models are necessary to determine local and global effects intrinsically linked with the three-dimensional pattern of the eddy currents, such as electromechanical loads on the structures or gap voltages across insulated elements.

As far as the mathematical model is considered, the transient electromagnetic phenomena can be described by the quasistationary Maxwell equations (5), (6), (7), and the constitutive equations (8), (9).

Of course, a correct and complete description of the problem should take into account not only the currents in the circuits and in the massive conducting structures, but also the electromagnetic behavior of the plasma. However, as it should be clear from the above discussion , electromagnetic models consistent with the MHD evolution of the plasma can hardly be employed in the presence of nonaxisymmetric geometries. This constraint is imposed by the formidable complexity of a fully 3-D consistent model and by the necessity to remain within reasonable limits of memory occupation and CPU time when carrying out numerical calculations.

Thus, if the attention is focused on the three-dimensional effects, it is necessary to accept the limitation that the plasma plays the role of a source term: the dynamics of the plasma current is assumed to be known either from experiments or two-dimensional simulations.

Even if the difficulty on the plasma side is overcome by assuming the plasma as a source term, the complexity of the geometries causes the Maxwell equations to be solved numerically. Several approaches are available to describe the eddy current problem; each of them has its own advantages and drawbacks. Thus, in the choice of the computational method, it is very useful to take into account the typical geometries and conditions encountered in a fusion reactor as well as the main objectives of the study.

The differences between the various formulations substantially consist in the selection of the primary unknown and in the nature of the spatial operators. A discussion of the different ways to formulate the eddy current problem is given in [13] and summarized in a schematic way in Table 1. Here we briefly recall that there exist two main families of formulations: those magnetic, based on the diffusion of the magnetic field H or the current density vector potential T, and those electric, related to the electric field E, its time integral $A^* = \int E \, dt$ or the magnetic vector potential A.

The magnetic formulations guarantee the solenoidality of the current density J, trying to impose the solenoidality of the flux density B in a weak form. Conversely, the electric approaches ensure the solenoidality of B but not that of J: as a result, the approximate solutions obtained using electric formulations may yield $J \cdot n \neq 0$ at the surface of the conductors.

As the principal interest is focused on the eddy currents in the structures, the magnetic formulations seem to be preferable for the electromagnetic analysis in fusion devices.

Another choice has to be made between integral and differential formulations. Integral formulations can hardly treat inhomogeneities or nonlinearities, and the calculation of the characteristic matrices may be highly time consuming. However, for these approaches, the discretization cost is lower than that required by differential formulations: for example, using volume integral formulations, only the conducting regions have to be discretized.

TABLE 1. Differential formulations of the eddy current problem.

METHOD	EQUATIONS		GAUGE
H - ψ	$\nabla \times (\sigma^{-1} \nabla \times H) = -\partial(\mu H)/\partial t$	in V_c	$(\nabla \cdot \mu H = 0)$ automatically
	$- \nabla \cdot (\mu \nabla\psi) = 0$	in $R^3 - V_c$	implied
T - Ω	$\nabla \times (\sigma^{-1} \nabla \times T) = -\partial[\mu(T-\nabla\Omega)]/\partial t$	in V	$\nabla \cdot T = 0$ or
	$- \nabla \cdot (\mu \nabla\Omega) = 0$	in $R^3 - V_c$	$T \cdot u = 0$ or ...
$A^* - \psi$	$\nabla \times (\mu^{-1}\nabla \times A^*) = - \sigma \partial(A^*)/\partial t$	in V_c	$(\nabla \cdot \sigma A^* = 0)$ automatically
	$- \nabla \cdot (\mu \nabla\psi) = 0$	in $R^3 - V_c$	implied
A - ϕ	$\nabla \times (\mu^{-1}\nabla \times A) = - \sigma (\partial A/\partial t + \nabla\phi)$	in V_c	$\nabla \cdot A = 0$ or
ψ	$- \nabla \cdot (\mu \nabla\psi) = 0$	in $R^3 - V_c$	$A \cdot u = 0$ or ...

For the next generation Tokamaks, the absence (or, at least, the very limited influence) of ferromagnetic materials allows for the choice of an integral formulation. Among these, boundary solution procedures are not so suitable because they can hardly handle resistivity inhomogeneities, whereas mixed formulations (differential methods coupled with boundary solution procedures at the surface of the conductors) may give rise to numerical difficulties, because the geometries of the conducting structures are characterized by the presence of thin layers and very narrow gaps across adjacent elements.

Thus, although mixed formulations can represent a valid possible alternative, the most attractive and simple approach seems to be a volume integral formulation in terms of the current density J. As far as the integral formulations are concerned, a number of different approaches based on the thin shell approximations were available. This simplification reduced the 3-D eddy current calculation to a 2-D problem on the shell surface. The solenoidality of the surface current density was assured either by introducing an additional unknown like the scalar electric potential [14, 15], or by selecting divergence-free basis vector functions as in the stream function [16, 17, 18] or in the finite element network approaches [19, 20]. These integral procedures are satisfactory to describe thin structures, even if some complications arise in the T joints.

However, the fully 3D pattern of the eddy currents in massive structures like blankets and shield modules called for the development of a fully 3D integral approach, applicable to describe, without any additional numerical overhead, situations in which both massive structures and thin plates are present.

4. AN INTEGRAL EDDY CURRENT FORMULATION BASED ON THE TWO-COMPONENT CURRENT DENSITY VECTOR POTENTIAL

The solenoidality of the flux density B can be ensured by introducing the magnetic vector potential A:

$$B = \nabla \times A \tag{37}$$

Obviously, A is not uniquely defined by this equation, since the addition of the gradient of an arbitrary scalar function does not modify its curl. Therefore, an additional constraint (gauge) must be imposed; for instance, the unicity can be assured by imposing:

$$\nabla \cdot A = 0 \tag{38}$$

with regularity conditions at infinity.

Eqs. (5) and (37) yield:

$$\nabla \times \left(E + \frac{\partial A}{\partial t} \right) = 0 \tag{39}$$

and hence

$$E = - \frac{\partial A}{\partial t} - \nabla \phi \tag{40}$$

where ϕ is the scalar electric potential.

As far as linear nonmagnetic conductors are concerned, using eqs. (6), (8) and (9), E and A can be expressed in terms of the unknown solenoidal current density J inside the conducting region V_c:

$$E(x, t) = \eta \, J(x, t) - E_{ext}(x, t) \tag{41}$$

$$A(x, t) = \frac{\mu_0}{4\pi} \int_{V_c} \frac{J(x', t)}{|x-x'|} \, dV' \tag{42}$$

where x stands for (x,y,z), x' for (x',y',z'), dV' for $dx'dy'dz'$, η is the resistivity, A is the magnetic vector potential due to the unknown current density J.

Combining eqs. (40), (41) and (42), the following integral formulation is obtained:

$$\eta \, J(x, t) + \frac{\mu_0}{4\pi} \int_{V_c} \frac{\partial J(x', t)/\partial t}{|x-x'|} \, dV' = E_{ext}(x, t) - \nabla\phi(x, t) \tag{43}$$

$$J \in V \tag{44}$$

$$J(x, 0) = J_0(x) \tag{45}$$

where J_0 is the prescribed initial condition and V is the set of vector fields defined in V_c which satisfy the following conditions:

$$\nabla \cdot v = 0 \quad \text{in } V_c \tag{46}$$

$$n \cdot v = 0 \quad \text{on } \partial V_c \tag{47}$$

For this formulation, the unicity of the solution for the current density J can be shown in the following way.

Let $(\delta J, \delta\phi)$ be the difference between two different possible solutions having the same initial conditions. Simple manipulations of eq.(43) yield:

$$\int_{V_c} \delta J \cdot \eta \delta J(x,t) dV + \frac{\mu_0}{8\pi} \frac{d}{dt} \int_{V_c} \int \frac{\delta J(x,t) \cdot \delta J(x',t)}{|x-x'|} \, dV \, dV' = -\int_{V_c} \delta J \cdot \nabla\phi(x, t) dV \tag{48}$$

where the r.h.s. is zero, due to conditions (46)-(47), and the two integrals on the l.h.s. are non negative quantities, being related to the expressions of the ohmic power and magnetic energy, respectively.

Hence:

$$\frac{\mu_0}{8\pi} \int_{V_c} \int_{V_c} \frac{\delta J(x,t) \cdot \delta J(x',t)}{|x-x'|} \, dV \, dV' = -\int_0^t \left[\int_{V_c} \delta J \cdot \eta \delta J(x,t) dV \right] dt \le 0 \tag{49}$$

The l.h.s. can be interpreted as the magnetic energy related to the current δJ; being this energy a non-negative quantity, it results to be zero, and then $\delta J=0$. The unicity of ϕ can be show similarly.

To obtain numerical solutions, J can be approximated as a linear combination of basis function J_k belonging to the set V:

$$J(x, t) = \sum_k I_k(t)J_k(x) \tag{50}$$

and the Galerkin method can be applied to eq. (43), yielding the following linear system [21, 22]:

$$\{L\} \frac{d[I]}{dt} + \{R\}[I] = [V] \tag{51}$$

where

$$L_{ik} = \frac{\mu_o}{4\pi} \int_{V_c} \int_{V_c} \frac{J_i(x) \cdot J_k(x')}{|x - x'|} \, dV \, dV' \tag{52}$$

$$R_{ik} = \int_{V_c} J_i(x) \cdot \eta \, J_k(x) \, dV \tag{53}$$

$$V_i = \int_{V_c} J_i(x) \cdot E_{ext}(x, t) \, dV \tag{54}$$

It should be noticed that using weighting functions belonging to V, the scalar electric potential ϕ does not appear in the calculation process. However, ϕ can be computed afterwards, by solving the following Neumann problem in V_c:

$$\nabla \cdot \eta \nabla \phi = \nabla \cdot \eta(E_{ext} - \partial A/\partial t) \qquad \text{in } V_c \tag{55}$$

$$\frac{\partial \phi}{\partial n} = n \cdot (E_{ext} - \partial A/\partial t) \qquad \text{on } \partial V_c \tag{56}$$

and the Dirichlet problem:

$$\nabla^2 \phi = 0 \qquad \text{in } R^3 - V_c \tag{57}$$

ϕ known on ∂V_c and regular at infinity $\tag{58}$

To verify (at least in a weak form) the condition of eq. (44) the shape functions J_k must satisfy the following relationships:

$$\oint_S J_k \cdot n \, dS = 0 \qquad \forall \text{ closed surface } S \tag{59}$$

$$n \cdot J_k = 0 \qquad \text{on } \partial V_c \tag{60}$$

The solenoidality condition (59) is assured by introducing a current density vector potential T defined by

$$\nabla \times T = J \tag{61}$$

and by the gauge [22, 23, 24, 25]

$$\mathbf{T} \cdot \mathbf{u} = 0 \tag{62}$$

where u is an arbitrary vector field which does not possess closed field lines.

As shown in [24, 25], the uniqueness of T, for a given current distribution, is assured by eq. (62). Let δT the difference between two different possible vector potentials giving the same current density J. It follows that $\nabla \times \delta T = 0$ in R^3 , and hence we can set $\delta T = \nabla \chi$. Let x_0 be a reference point arbitrarily fixed and x be any point in R^3. If it is always possible to find a field line Γ of u which connects x_0 to x, then we have:

$$\chi(\mathbf{x}) = \chi(\mathbf{x}_0) + \int_\Gamma \nabla \chi \cdot d\mathbf{l} = \chi(\mathbf{x}_0) \tag{63}$$

because $d\mathbf{l} \parallel \mathbf{u}$ and $\nabla \chi \cdot \mathbf{u} = 0$.

Therefore, provided that the field lines of u are able to connect any point x with the reference point x_0, it follows that χ is constant and hence $\delta T = \nabla \chi = 0$ in R^3. A simple example is given by $\mathbf{u}(\mathbf{x}) = \mathbf{x}$ with x_0 selected in the origin.

It should be noticed that the vector field u has to verify an additional requirement, i.e. it must not possess closed field lines. In fact, a zero circulation of T along a closed field line would imply that the linked current flux is forced to be zero.

Finally, we wish to note that the vector field u is not necessary continuous.

After introducing the two-component current density vector potential T, the shape functions for J can be deduced from a set of basis functions for T, and condition (62) allow us to limit the number of scalar unknowns to the two nonzero components of T.

Notice, however, that this approach, when used in conjunction with standard isoparametric elements, cannot adequately take into account the resistive discontinuities [13]. This is shown by the following example: consider a coordinate system in which $\mathbf{u} = \mathbf{y}$; suppose that there is a resistive discontinuity across the plane $x = $ constant; then, not only the normal component $J_x = \partial T_z / \partial y$ is continuous, but also $J_z = -\partial T_x / \partial y$ is forced to be continuous; as a consequence, the tangential component $E_z = \eta J_z$ has a jump.

The continuity of the tangential component of T across adjacent elements is by itself sufficient to guarantee the continuity of the normal component of J, involved by the solenoidality condition (59); therefore, the continuity of the normal component of T is not required. This suggests the possibility of employing the edge elements, proposed by Nedelec [26] and Bossavit [27] and generalized by van Welij [28] and Kameari [29], whose degrees of freedom are the path integrals (or, equivalently, the tangential components) of the unknown vector field along the edges. Clearly these elements guarantee continuity of the tangential components across element boundaries.

The adoption of this finite element approximation allows for a very attractive way of imposing gauge condition (62) from a numerical point of view [22, 25, 30]. Consider the graph made up of nodes and edges of the finite element mesh approximating the integration domain. The discrete analog of condition (62) is obtained by identifying the direction of u along an arbitrary tree of the graph of the mesh. In this way, all the nodes are connected and no closed loops are formed by definition. As a consequence of this choice, the degrees of freedom corresponding with the edges of the tree are eliminated. On the other hand, the remaining degrees of freedom correspond to the fluxes of $J=\nabla\times T$ linked with the set of independent loops closed by the tree and each of the residual edges (i.e. the edges of the co-tree).

The shape functions defined in this way automatically verify eq. (59), but not eq. (60). However, this condition can easily be imposed if the domain is simply connected. If the tree is built by connecting first the subset of the nodes of ∂V_c with boundary edges only, then every degree of freedom associated with a boundary edge of the co-tree represents a net current flux crossing a surface which lies on the boundary. Therefore it must be suppressed. Numerically, the boundary part of the tree can readily be formed as the first part of the tree, by means of the sequential algorithm described in [31]. This procedure corresponds to impose the boundary condition:

$$T \times n = 0 \qquad \text{on } \partial V_c \qquad (64)$$

which implies eq. (60), and is equivalent to it when the vector field u is selected such as

$$u \cdot n = 0 \qquad \text{on } \partial V_c \qquad (65)$$

The elimination of these boundary degrees of freedom can also be explained by Euler's formula for a polyhedral surface [32]:

$$V - E + F = 2 - 2p \qquad (66)$$

where V is the number of vertices (i.e. the nodes), E the number of edges, F the number of faces, and p the genus of the polyhedron.

The number N_d of degrees of freedom belonging to the boundary (left after imposing the gauge condition) is given by the number of edges belonging to the co-tree of the boundary graph. Therefore $N_d=E-(V-1)$, being $V-1$ the number of edges of the boundary tree. On the other hand, one has to impose zero current flux across each elementary face. Thus an homogeneous system of N_e independent linear equations is obtained, being $N_e = F-1$, because of the solenoidality of $\nabla \times T$.

If the region is simply connected, then $p=0$ and $N_d = N_e$. Therefore the trivial solution is also the unique one.

Conversely, if $p>0$, there are $2p$ independent distributions T_{add} on the boundary which give $J \cdot n=0$. However only p distributions are related to net current fluxes circulating inside the conducting region, whereas the other p ones would be related to currents circulating in R^3-V_c and give rise to current distributions inside V_c which can be obtained also as linear combinations of the other degrees of freedom. Therefore, the correct additional degrees of freedom can be chosen by minimizing the quadratic form:

$$\Xi = \frac{1}{2} \int_{V_c} |\nabla \times T - \nabla \times T_{add}|^2 dV \tag{67}$$

where T is approximated as a linear combination of the shape functions. If, as a result of the minimization procedure, we find a solution giving $\Xi = 0$, then T_{add} is redundant. On the other hand, if $\Xi > 0$, the distribution related to T_{add} cannot be reproduced by internal degrees of freedom and should therefore be included among the shape functions. The procedure goes on until the p independent additional basis functions T_{add} are found, allowing for the automatic pre-processing of multiply connected domains [33].

The mesh, the tree, the co-tree and the active edges are shown in Fig. 4 and Fig. 5, with reference to a simply and multiply connected domain, respectively.

The structures of interest have an high level of symmetry, related to the toroidal shape of the device; it is important, in order to save computer time and memory occupation, to have a correct and efficient model of the symmetries involved.

Two classes of symmetry can be defined with reference to a toroidal component of a Tokamak, like the vacuum vessel or the first wall. One is the symmetry of reflection with respect to a poloidal plane and/or to the equatorial plane; the other one is the symmetry with respect to a given rotation around the axis of the torus.

The basic idea to take into account these symmetries is to assume basis functions J_k which automatically verify the symmetry conditions. For this purpose the integration domain V_c to be discretized is an elementary part of the whole structure. The shape functions in the rest of the structure can be obtained by means of suitable operators.

For instance, with reference to a system of rectangular coordinates with unit vectors t_x, t_y and n_z, with n_z perpendicular to the plane of symmetry, we define the reflection operator:

$$\{S_n\} = \begin{bmatrix} 1 & 0 & 0 \\ 0 & 1 & 0 \\ 0 & 0 & -1 \end{bmatrix} \tag{68}$$

such that the components of J_k at the reflected point $[x_r] = \{S_n\} [x]$ are given by:

$$[J_k(x_r)] = \epsilon \{S_n\} [J_k(x)] \tag{69}$$

where $[x] = (x \cdot t_x, x \cdot t_y, x \cdot n_z) = (x, y, z)$ is in the integration domain V_c, $[x_r]$ is in the reflected domain, $[J_k]$ stands for $(J_k \cdot t_x, J_k \cdot t_y, J_k \cdot n_z)$, and ϵ is +1 or -1, depending on the type of the symmetry.

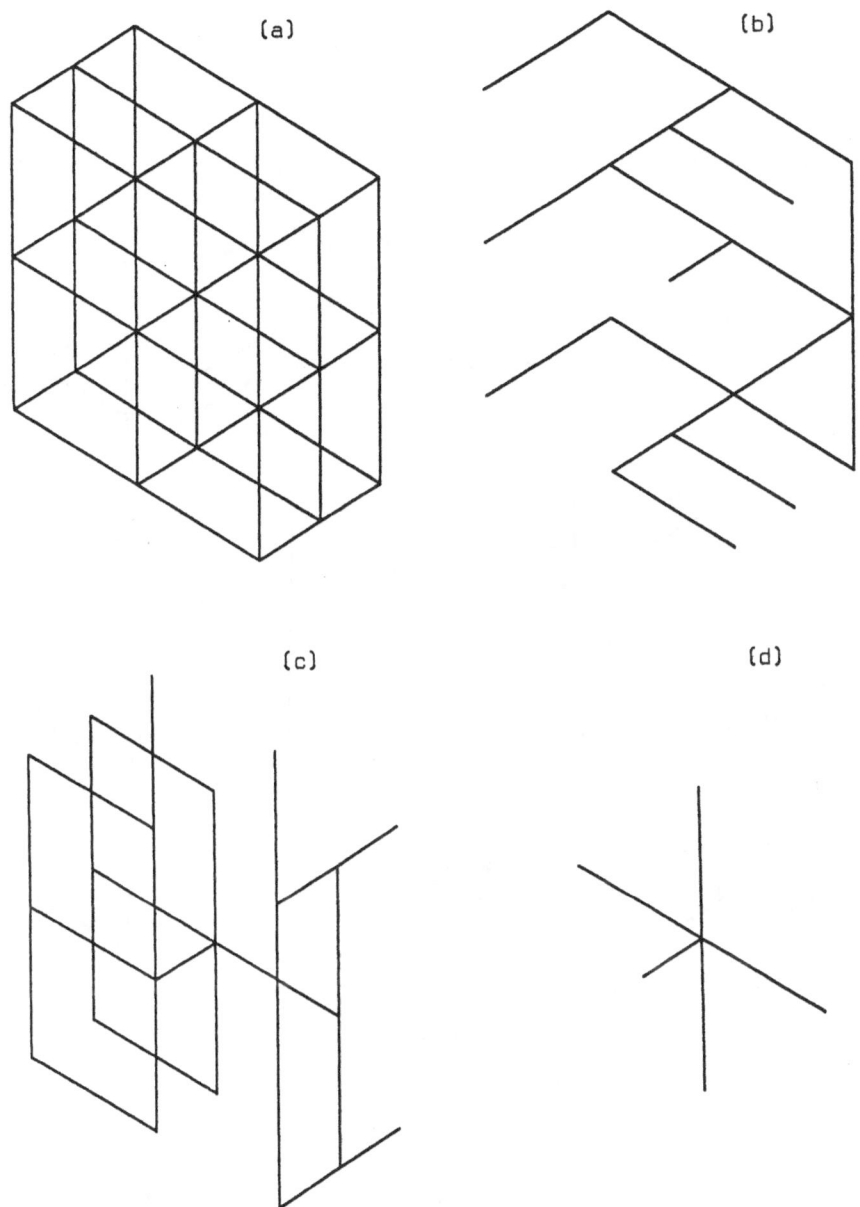

Figure 4. Conducting brick (simply connected region): a) mesh; b) tree;
c) co-tree; d) active edges.

182

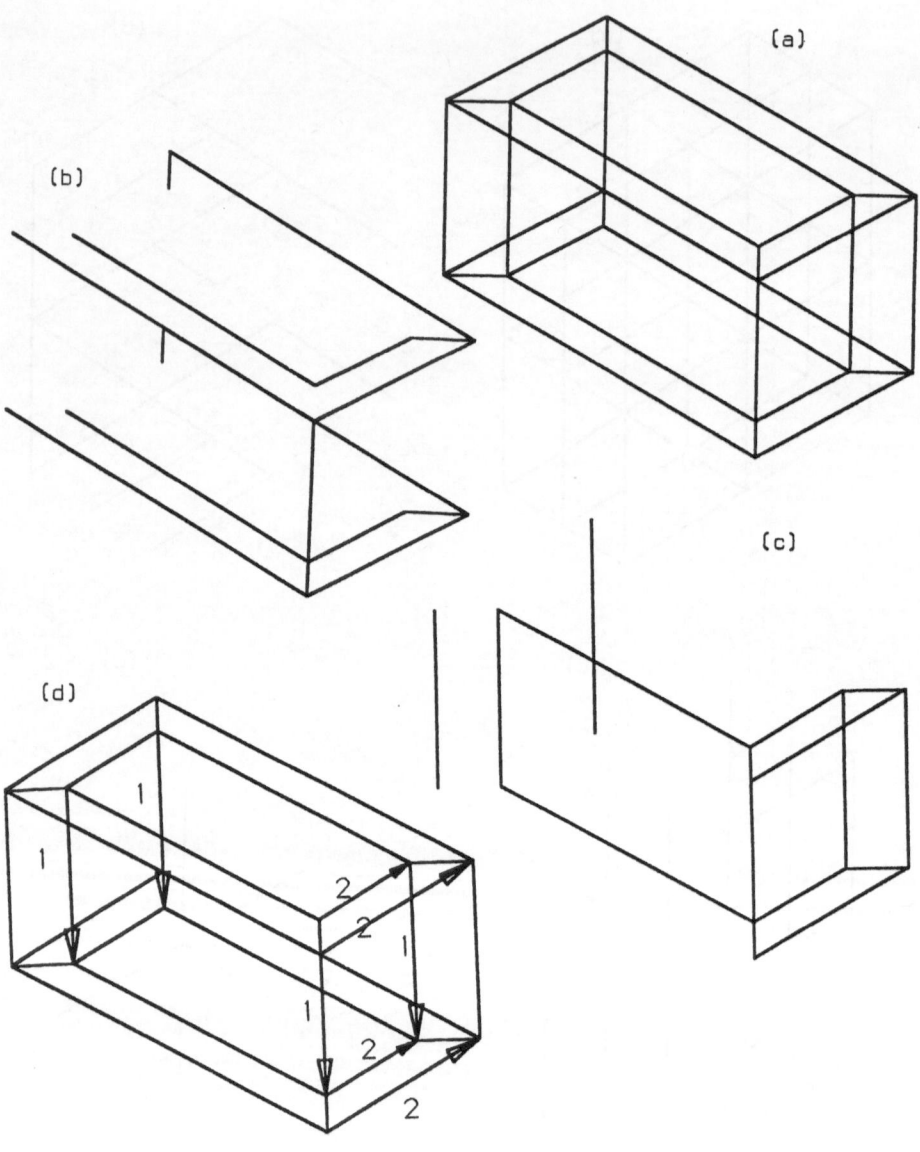

Figure 5. Short rectangular solenoid (multiply connected region):
a)mesh; b) tree; c) co-tree; d) active edges (only the
degree of freedom n.1 is really active).

The rotation of an angle α around the x axis (the axis of the torus) can be represented by the operator of rotation $\{R_\alpha\}$:

$$\{R_\alpha\} = \begin{bmatrix} 1 & 0 & 0 \\ 0 & \cos\alpha & -\sin\alpha \\ 0 & \sin\alpha & \cos\alpha \end{bmatrix} \tag{70}$$

such that the components of J_k at the rotated point $[x_r]=\{R_\alpha\}$ $[x]$ are given by:

$$[J_k(x_r)] = \{R_\alpha\}\ [J_k(x)] \tag{71}$$

It should be noticed that the continuity of $J\cdot n$ has to be assured on the symmetry planes. Therefore, for instance, when applying a reflection with $\epsilon=+1$, the condition $J\cdot n=0$ must be guaranteed on the symmetry plane. This can directly be imposed on the shape functions using the standard procedure applied to impose the boundary condition (60).

At this point, the calculation of the coefficients of $\{L\}$, $\{R\}$ and $[V]$ is straightforward. For instance, with reference to the part of the vacuum vessel shown in Fig. 6, where a symmetry of reflection ($\epsilon=-1$) with respect to the plane z=0 and a modularity of order 16 in the toroidal direction are implied, the terms of $\{L\}$, $\{R\}$ and $[V]$ have the following form:

$$L_{ik} = 2N\alpha \sum_m^{N\alpha} \sum_j^{1} \frac{\mu_0}{4\pi} \int_{V_c}\int_{V_c} \frac{[J_i(x)]\{R_\alpha\}^m \epsilon^j \{S_n\}^j [J_k(x')]}{|x-\{R_\alpha\}^m \{S_n\}^j [x']|} \, dV \, dV' \tag{72}$$

$$R_{ik} = 2N\alpha \int_{V_c} J_i(x)\cdot \eta\ J_k(x)\ dV \tag{73}$$

$$V_i = 2N\alpha \int_{V_c} J_i(x)\ \cdot\ E_{ext}(x, t)\ dV \tag{74}$$

Here $\{R_\alpha\}^m$ represents the operator which imposes a rotation $m\alpha$ around x, $N\alpha = \dfrac{2\pi}{\alpha}$ the number of modules around x.

The evaluation of the effects of a transient electromagnetic phenomenon requires some more steps after the computation of the eddy current density distribution. Depending on the particular design phase, it is important to know total currents flowing through specific cross-sections of the conducting structure, magnetic fields and fluxes at given points, related electromagnetic force distributions. In the integral formulation here described, the total currents flowing through given surfaces made out of faces of elements are linear combinations of the primary unknowns of the problem, i.e. the path integrals of T along the active edges of the finite element domain. Magnetic fields can be computed by integrating the Biot-Savart law over each element of the domain. In particular, in order to save computer time, it is useful to define a rectangular matrix relating the field at given points to the degrees of freedom describing the eddy current density distribution. Contributions to the field produced by external currents are computed by means of numerical or, when possible, analytic formulae. The forces are finally given by integrating the volume force density $J \times B$.

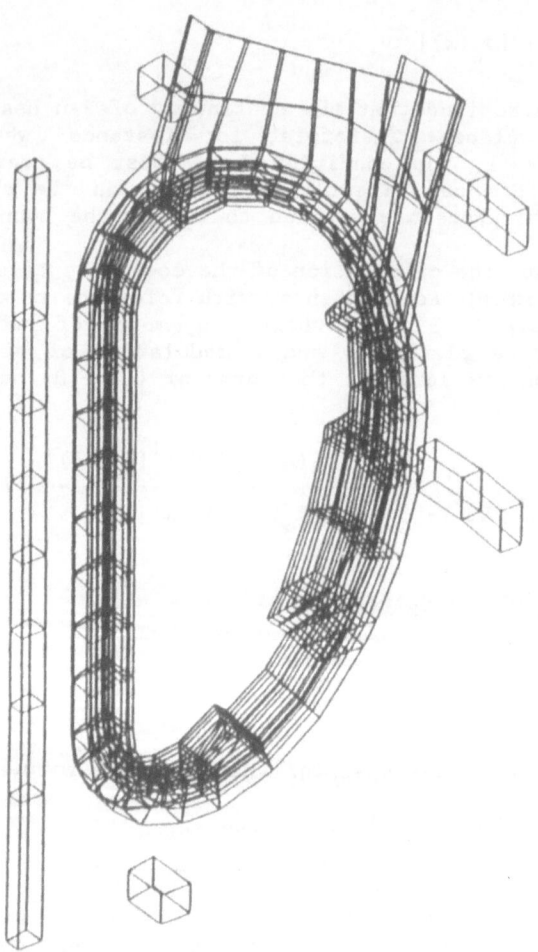

Figure 6. Schematization of the NET vacuum vessel and PF coil system. Only $\frac{1}{2} \frac{1}{16}$ of the structure has been discretized.

Coming to the conclusions, it has been shown how an integral formulation using edge elements can give a satisfactory solution to the problem of the computation of the eddy currents in non-magnetic structures. In particular, this formulation is especially attractive for the analysis of magnetic fusion devices, where both massive structures and thin shells are present, and the electrical resistivity is often discontinuous. It is interesting to see that, in the limit of thin structures, all the active edges are internal and can be directed along n (the unit vector normal to the shell of thickness Δ)if hexahedral elements are used. Therefore the unknowns (the path integrals of T along these edges) exactly correspond to the values of the surface current stream function $\psi=T\Delta$, defining the surface current density J_s as:

$$J_s = J \Delta = \nabla \times T\Delta n = \nabla\psi \times n \qquad (75)$$

In this way, the integral formulation exactly limits to the one studied in [16, 17, 18].

However, the formulations directly based on ψ cannot handle complex topologies like T joints in a simple way. Conversely, the one based on hexahedral edge elements automatically assures the solenoidality of J in any case, without any additional numerical overhead, and is also able to model massive 3D structures coexisting with thin plates.

A particularly delicate aspect of the integral methods is related to the computation of the {L} matrix, which is in general expensive and difficult in the proximity of the $1/|x\text{-}x'|$ singularity. A method of computing the singular contribution is discussed in [22]. On the other hand, the integral methods are particularly suited for parallel computing [34]. Furthermore, once the {L} matrix has been computed for a given mesh, several studies with different forcing terms and conducting materials can be analyzed using a very limited additional amount of CPU time.

Finally, we would like to stress that the integral formulations in terms of current density shape functions give rise to numerical models which can easily be interpreted as that of a lumped parameter L-R network. In this way, it is particularly easy to couple the eddy current model to other phenomena described in terms of lumped parameter circuits like, for instance, in the case of the analysis of plasma vertical instability [35, 36].

5. REFERENCES

1. Keilhacher, M. and the JET Team (1990) 'The Jet H-Mode at High Current and Power levels', Proc. of the 12th International Conference on Plasma Physics and Controlled Thermonuclear Research, Nice (France), October 1988, Nuclear Fusion, IAEA, suppl.
2. Freidberg, J.P. (1987) 'Ideal Magnetohydrodynamics', Plenum Press, New York
3. Grad, H., Rubin, H. (1958) 'Hydromagnetic Equilibria and Force-Free

Fields' in Proceedings of the Second United Nations International Conference on the Peaceful Uses of Atomic Energy, United Nations, Geneva, Vol. 31, 190-197

4. Shafranov, V.D. (1958) 'On magnetohydrodynamical equilibrium configurations', Sov. Phys.-JETP, Vol 6 (33), Number 3, 545-554

5. Coccorese, E., Martone, R. (1978) 'A design procedure for air core transformers in toroidal axisymmetrical geometry', Proc. X SOFT Symposium on Fusion Technology, Padua, Ed. Pergamon Press.

6. Albanese, R. (1986) 'Time Evolution of a Magnetically Confined Plasma', PhD thesis, (Dept El. Eng., University of Naples, in italian)

7. Blum, J., Le Foll, J. (1984) 'Plasma Equilibrium Evolution at the Resistive Diffusion Time Scale', Computer Physics Reports 1, 465-494

8. Jardin, S.C., Pomphrey, N., De Lucia, J. (1986) 'Dynamic modeling of transport and positional Control of tokamaks', Jour. Comp. Phys., 66, 481

9. Albanese, R., Blum, J., De Barbieri, O. (1987) 'Numerical studies of the Next European Torus via the PROTEUS Code', 12th Conf.on the Num. Symulation of Plasmas, S. Francisco

10. Grad, H., Hu, P.N., Stevens, D.C. (1975) 'Adiabatic evolution of plasma equilibrium', Proc. US Nat. Acad. Science, 72, 3789-3793.

11. Albanese, R, (1989) 'Analysis of the Plasma Equilibrium Evolution in the Presence of Circuits and Massive Conducting Structures', Proceedings of XV SOFT Symposium on Fusion Technology, 281-286, Elsever Science Publishers B.V.

12. Coccorese, E., Coco, S., Martone, R. (1990) 'A model reduction technique for high order eddy current problems', IEEE Trans. MAG-26, 690-693

13. Albanese, R., Rubinacci, G. (1990) ' Formulation of the eddy current problem', IEE Proc. A, 137, 16-22

14. Kobayashi, T. (1979) 'Analysis of eddy currents induced on the vacuum vessel of a tokamak', Jpn J. Appl. Phys. 18, 2003-2009

15. Rubinacci, G. (1983) 'Numerical computation of eddy currents on the vacuum vessel of a tokamak', IEEE Trans MAG-19, 2478-2481

16. Kameari, A. (1981) 'Transient eddy current analysis in thin conductors with arbitrary connections and shapes', J. Comput. Phys. 42, 124-140.

17. Bossavit, A. (1981) 'On the numerical analysis of eddy current problems', Comput. Methods Appl. Mech.& Eng. 27, 303-318

18. Blum, J., Dupas, L., Leloup, C., Thooris, B. (1983) ' Eddy current calculations for the Torus Supra tokamak', IEEE Trans. MAG-19, 2461-2464

19. Weissenburger, D.W., Christensen, U.R. (1982) 'A network mesh method to calculate eddy currents on conducting surfaces', IEEE Trans. MAG-18, 422-425

20. Turner, L.R., Lari, R.J. (1982) 'Applications and further developments of the eddy current program Eddynet', IEEE Trans. MAG-18, 416-421

21. Albanese, R., Martone, R., Miano, G., Rubinacci, G. (1985) 'A T formulation for 3-D finite element eddy current computation', IEEE

Trans. MAG-21, 2299-2302

22. Albanese, R., Rubinacci, G. (1988) 'Integral formulation for 3D eddy current computation using edge elements', IEE Proc. A, 135, 457-462

23. Carpenter, C.J. (1977) 'Comparison of alternative formulations of 3-D magnetic field and eddy current problem at power frequencies', IEE Proc., 124, 1026-1034

24. Brown, M.L. (1982) ' Calculation of 3-dimensional eddy currents at power frequencies', IEE Proc. A, 129, 46-53

25. Albanese, R., Rubinacci, G. (1990) 'Magnetostatic field computations in terms of two-component vector potentials', Int. J. Num. Meth. Eng., 29, 515-532

26. Nedelec, J.C. (1980) 'Mixed finite elements in R^3', Numer. Math., 35, 315-341

27. Bossavit, A. (1988) 'A rationale for edge elements in 3-D fields computations', IEEE Trans. MAG-24, 74-79

28. Van Welij, J.S. (1985) 'Calculation of eddy currents in terms of H on hexahedra', IEEE Trans. MAG-21, 2239-2241

29. Kameari, A. (1989) 'Calculation of transient 3D eddy current using edge elements', IEEE Trans. MAG-26, 466-469

30. Albanese, R., Rubinacci, G. (1988) 'Solution of three dimensional eddy current problems by integral and differential methods', IEEE Trans. MAG-24, 98-101

31. Hale, H.W. (1961) 'A logic for identifying the trees of a graph', Trans AIEE PAS-80, 195-198

32. Griffiths, H.B., Hilton, P.J., (1970) 'A comprehensive text of classical mathematics. A contemporary interpretation', Van Rostrand Rehinold, London

33. Albanese, R., Rubinacci, G. (1989) 'Treatment of multiply connected regions in two-component electric vector potential formulations', IEEE Trans. MAG-26, 650-653

34. Bryant, C.F., Roberts, M.H., Trowbridge, C.W., (1989)'Implementing a boundary integral method on a transputer system', IEEE Trans. MAG-26, 819-822

35. Albanese, R., Coccorese, E., Rubinacci, G., (1989)'Plasma modeling for vertical instabilities', Nuclear Fusion, 29, 1013-1023

36. Albanese, R., Coccorese, E., Martone, R., Rubinacci, G., (1989) 'Analysis of vertical instabilities in air core tokamaks in the presence of three dimensional conducting structures', IEEE Trans. MAG-26, 853-856.

ELECTROMAGNETIC PROBLEMS IN FUSION REACTOR DESIGN

E. COCCORESE
Facoltà di Ingegneria
Università di Reggio Calabria
via Veneto 69
I-89127 Reggio Calabria - Italy

and

R. MARTONE
Istituto di Ingegneria Elettronica
Università di Salerno
I-84081 Baronissi (Salerno)
Italy

ABSTRACT. The class of electromagnetic phenomena occurring in a fusion reactor based on the tokamak concept is so wide that a variety of models and related numerical techniques are needed to give a satisfactory solution to the main design problems.

Magnetostatics both in 2D and in 3D geometry is the basic methodology for the design of the magnetic configuration required by the tokamak concept: typical procedures and examples are illustrated in the second section.

A lumped parameter approach is more suited to tackle the problem of the control of the magnetic configuration, as discussed in the third section.

The design of the metallic structure of a tokamak reactor is strongly affected by plasma disruption events: an eddy current formulation is the most suited to this problem, as illustrated in the last section.

The examples reported refer to studies for next generation large tokamaks.

1. Introduction

Electromagnetic phenomena play a major role in fusion devices based on magnetic confinement; in this respect it suffices to say that the tokamak concept is based on the interaction between the current flowing in the plasma and a magnetic field of suitable configuration and strength.

In a broad sense, electromagnetic effects and phenomena occurring in a tokamak reactor cover with a surprising continuity the whole frequency spectrum, from static fields up to X and γ rays. However, the dynamics of the phenomena which have a more relevant impact on the engineering design of the device is such that electromagnetic propagation effects can be neglected; therefore we limit here ourselves to the treatment of electromagnetic problems which can be formulated in terms of static or quasi-static Maxwell equations.

In spite of this limitation, the class of electromagnetic phenomena occurring in a fusion reactor is still so wide that a variety of models and related numerical techniques are required for a satisfactory and safe treatment of the problems. Methodologies differ substantially from each other not only as far as their specific focus of interest is concerned (e.g. external circuits, eddy currents, plasma), but also as regards the primary goals of the investigation (e.g. selection of design guidelines, accurate computation of electromechanical forces, simulation of the plasma behaviour).

With respect to present day devices, in next generation tokamaks design problems are more severe because of the increased complexity of the structures and the upgrading of the plasma current. In addition, the presence of superconducting coils poses new problems such as the

189

Y. R. Crutzen et al. (eds.), Industrial Application of Electromagnetic Computer Codes, 189–214.

minimization of the a.c. losses produced by time-varying fields. For this reason, the development of new computation tools has been strongly encouraged and it is still under way.

From the electromagnetic point of view, methods and associated numerical procedures fall into three main categories:
a) magnetostatic models, which are particularly well suited for the design of the magnetic configuration;
b) lumped parameter models, which are particularly well suited for the design of the plasma control system;
c) eddy current models, which are particularly well suited for the study of the electromagnetic effects of plasma disruptions.

Obviously, all models must take into account that the plasma, being a conducting fluid, is not a purely electromagnetic object; for this reason, standard or even commercial electromagnetic codes are usually not directly applicable to the problems of interest.

In the following sections, it is aimed to illustrate some typical electromagnetic problems and how they can be solved with the help of procedures belonging to the above mentioned categories; the examples reported refer to studies for next generation large tokamaks.

2. Design of the magnetic configuration

The magnetic configuration in a tokamak device [1] is created by two different systems of coils:
a) the Toroidal Field (TF) coils;
b) the Poloidal Field (PF) coils.

The TF coils have the task to create a sufficiently strong magnetic field over the plasma region with a field ripple to be kept within acceptable values .

The PF coils provide the necessary field to create and maintain the plasma current throughout the operating cycle.

The poloidal field can be split into an equilibrium part, which controls the position and the shape of the plasma, and a part which drives the plasma current having a negligible field in the plasma region.

Fig. 1 gives a schematic picture of the TF and PF coil arrangement.

For the future fusion reactors, there is a tendency to use superconductors for both TF and PF coils, to keep the electric power within acceptable values.

With superconducting TF coils, the only task of the power supply system is to energize the coils (at low power) before the operation.

On the contrary, even with superconducting coils, the power required to supply the PF coil system is much higher due to the variability of the PF currents during the operating cycle.

A significant part of this power is driven in feedback by an on-line detection system of the plasma parameters.

2.1. TF SYSTEM

The design of the TF coils should take into account the following aspects:
a) coil geometry: taking into account the constraints coming from the plasma geometry (i.e. plasma major and minor radius, elongation, etc.) and from the presence of other machine components (i.e. first wall, blanket, vacuum vessel, etc.) the coil shape is chosen minimizing the magnetic energy and the mechanical stress.;
b) number of coils: this should be chosen as a trade-off between the two contrasting requirements

of sufficiently large access ports and low field ripple;

c) field limitation: when using superconducting coils, the present technology imposes a limit on the maximum field in the superconductor.

Design problems of non-electromagnetic nature are clearly beyond the scope of this paper; electromechanical and electrodynamic problems will be presented in the following sections. We now limit ourselves to some purely magnetostatic aspects, whose treatment require the use of 3D static codes.

These kinds of codes can be used for the computation of the field map produced by the current flowing in the coils; in particular one can be interested to compute the field ripple in the plasma region.

In case ferromagnetic materials are present one obviously needs a non linear code. This is, for example, the case of ferromagnetic inserts located underneath the TF coils to reduce the ripple [2].

Fig. 2 shows how the ferromagnetic material reduces the ripple by offsetting the inward push of the field lines from the TF coil.

2.2. PF SYSTEM

The design of the PF system depends strongly on the plasma characteristics.

A complete treatment of the plasma would imply the consideration of its fluidodynamic nature in terms of pressure, density, temperature of the various species of particles.

A simpler and more handy model is the so called MHD model [3, 4] where the plasma is treated as a continuous fluid subjected to electromagnetic and pressure forces.

For a number of engineering applications it is useful to treat the plasma a purely electromagnetic object; this leads to an even simpler model based on the circuit theory.

The following part of this section shall be devoted to the static aspects while dynamic ones will be treated in the next sections.

2.2.1. *Equilibrium Winding.* In the framework of the ideal MHD theory, the equilibrium condition for the plasma is expressed by the equation:

$$\mathbf{J} \times \mathbf{B} = \text{grad } p \tag{1}$$

where:

$\mathbf{J}$ and $\mathbf{B}$ are the current density and the magnetic field in the plasma, satisfying the Maxwell equations;

p is the kinetic pressure in the plasma.

In cylindrical coordinates, for an axisymmetric system like the tokamak, eq. (1) can be rewritten as [5, 6]:

$$\left(\frac{\partial^2}{\partial r^2} - \frac{1}{r}\frac{\partial}{\partial r} + \frac{\partial^2}{\partial z^2}\right) \psi(r,z) = -2\pi\mu_0[r^2 A(\psi) + B(\psi)] \tag{2}$$

where:

ψ is the flux function of the poloidal magnetic field;

$A(\psi)$ and $B(\psi)$ depend on the profiles of the pressure p and of the poloidal current I, as:

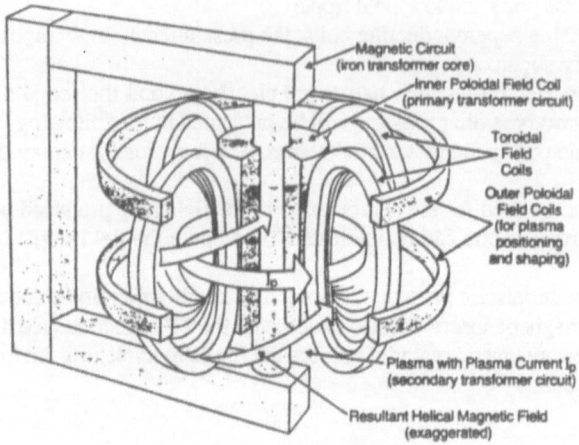

Fig. 1 The magnetic configuration in a tokamak.

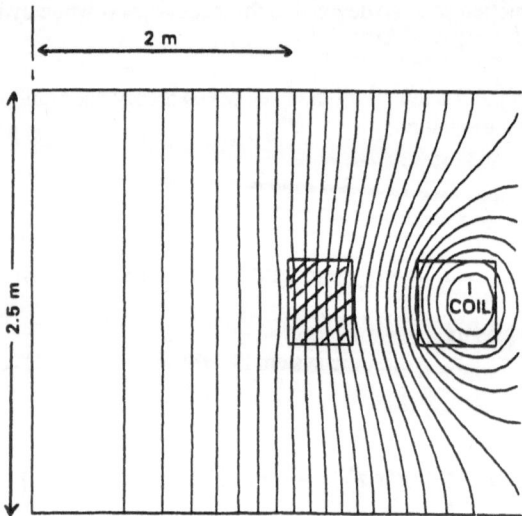

Fig. 2 Example of ripple correction with ferromagnetic inserts.

$$A(\psi) = 2\pi \frac{dp}{d\psi} \qquad\qquad B(\psi) = \frac{\mu_0}{4\pi} \frac{dI^2}{d\psi}$$

Two typical problems can be formulated for eq. (2):

i) the Fixed Boundary problem which corresponds to assigning the plasma boundary and the value of ψ on it [7];

ii) the Free Boundary problem which corresponds to assigning the external currents required to maintain the plasma ring in equilibrium without a-priori specifying the boundary of the plasma [8].

From an engineering point of view these two problems can be interpreted respectively as follows:

i) design of the poloidal configuration where one is faced with the problem of choosing a suitable set of external conductors able to produce the desired plasma shape;

ii) analysis of the poloidal configuration, where the unknown of the problem is the plasma equilibrium determined by a given set of external conductors.

A number of codes are available for the solution of eq. (2) for both problems. For example the technique described in [9,10] has been extensively used for NET design [11] as shown in fig. 3.

Each individual equilibrium configuration determines uniquely the magnetic field in the plasma region. This can be regarded as the sum of two contributions:

i) the field produced by the toroidal current flowing in the plasma;

ii) the external maintaining field to be produced by a suitable set of currents flowing in the PF coils.

Obviously the latter field can be created in infinite ways depending on number, location and distribution of the current in the PF coils.

The designer can exploit this degree of freedom by selecting among the infinite solutions the optimum one with respect to some index of quality.

2.2.2. *Transformer Winding*. Even if for the future reactors there is a belief that the toroidal current in the plasma can be driven with non-inductive mechanisms [12], present day -as well as next generation- tokamaks are based on an inductive drive of the plasma current.

The transformer winding has then the duty to provide the flux swing needed to induce and maintain the required plasma current.

To avoid any interference with the plasma equilibrium, the current flowing in the transformer winding should produce a negligible magnetic field in the plasma region. This is achieved by suitably choosing the geometrical distribution of the transformer current.

The flux swing to be produced by the transformer is given by :

$$\Delta\psi_T \quad = \quad M\Delta I_T \quad = \quad L_P I_P + \psi_{ohm} + \Delta\psi_E \tag{3}$$

where:

M is the mutual inductance between the plasma and the transformer winding;

ΔI_T is the variation of the ampere-turns in the transformer;

L_P is the plasma self inductance;

I_P is the plasma current;

ψ_{ohm} is the ohmic flux due to the non-zero resistivity of the plasma;

ψ_E is the contribution of the currents flowing in the equilibrium winding.

Eq. (3) shows how, for a given geometry, the transformer imposes a limits both on the maximum plasma current and on the duration of the pulse; this is due on practical limitation on the ampere-turns in the transformer. In any case, a minimization of the ampere-turns is achieved with a symmetric current swing, i.e. the current in the transformer is reversed.

Eq. (3) also shows how the contribution of the equilibrium winding can relieve the transformer duty. This suggests the idea of not having two separate windings for equilibrium and transformer actions. As a matter of fact one could design a comprehensive PF coil system capable to not only generate the magnetic field to hold the plasma in equilibrium but also generate the electric field to build up the plasma current and maintain it through a burn cycle.

This design philosophy has been applied for example in NET [11] whose PF coil arrangement is shown in Fig.4. Actually the NET machine uses an efficient solution for PF coils by having a double equilibrium and flux generation function for each coil set. For a typical scenario the PF coils begin at a positive current in the coils P1A, P1B, P1C and P1D (central solenoid) before swinging through zero to a negative current at the end of the pulse; these generate then the plasma current and resistive flux. At the same time the P3 and P4 coils begin at zero current and are ramped up nearly proportionally to the plasma current; these generate then the external maintaining field and contribute to the flux swing. Therefore each individual coil has a main duty and a secondary function which is matter of optimization of the overall system. For these reasons the six coils are separately supplied and controlled throughout the cycle.

In the following, we describe the numerical procedure used for the design of the NET PF configurations.

2.3.   DESIGN PROCEDURE FOR PF COIL CONFIGURATION

The procedure consists in the following steps:
i)   static MHD equilibrium
       input: - desired plasma parameters and shape
       output: - poloidal flux ψ_p produced by the plasma current on a discrete set of points
                belonging to its boundary;
ii)   PF coil design
       input: - poloidal flux ψ_p from step i)
       input: - permissible region for the location of the PF coils
                - maximum current density in the coils
                - ohmic flux ψ_{ohm}
                - tolerance for the the fulfilment of the uniformity condition of the flux on the plasma
                  boundary
       output: - location and currents of the PF coils;
iii) static MHD equilibrium
     input: - currents flowing in the PF coils
                 - global plasma parameters
     output: - plasma shape and parameters.

The procedure is iterated until a satisfactory matching between the desired and the obtained configuration is achieved; if necessary, constraints and tolerances should be relaxed.

Steps i) and iii) are performed with the aid of a static MHD equilibrium code, respectively using a fixed and free boundary formulation.

Step ii) is realized taking advantage of a linear programming formulation of the problem, which

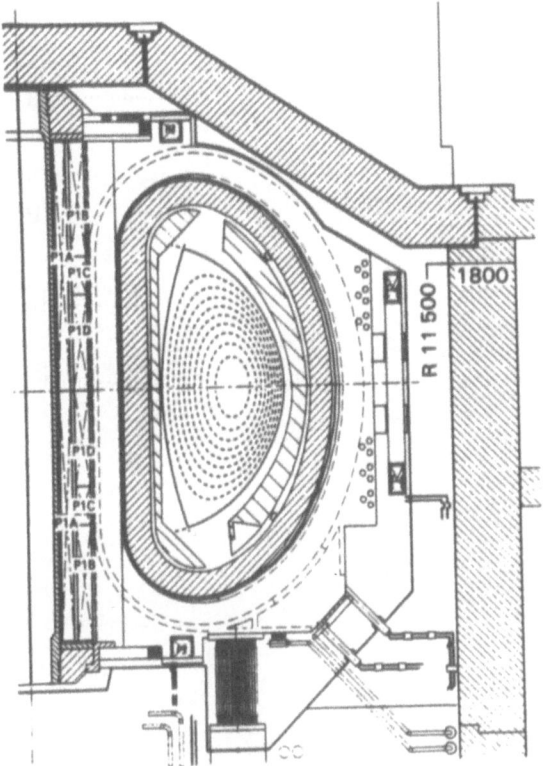

Fig. 3   A double null equilibrium configuration in NET

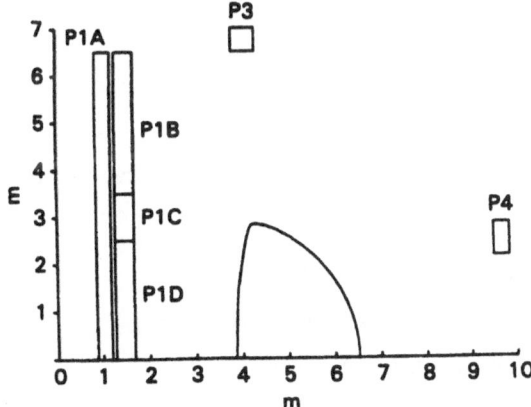

Fig. 4   PF coil configuration in NET

can be summarized as follows:

$$\psi_B \, (1-\varepsilon_E) \leq \sum_{j=1}^{N_E} M^E_{ij} \, I^E_j + \psi_P \, (r_i, z_i) \leq \psi_B \, (1+\varepsilon_E)) \qquad\qquad i=1,2,.........N_B$$

$$\psi_T \, (1-\varepsilon_T) \leq \sum_{j=1}^{N_T} M^T_{ij} \, I^T_j \leq \psi_T \, (1+\varepsilon_T)) \qquad\qquad i=1,2,.........N_B$$

$$|I^E_j| \leq G_E \, s^E_j \qquad\qquad\qquad j=1,2,.........N_E$$

$$|I^T_j| \leq G_T \, s^T_j \qquad\qquad\qquad j=1,2,.........N_T$$

$$v^E + v^T = min$$

with:

$$\psi_B = \psi_E \, (r_i, z_i) + \psi_P \, (r_i, z_i)$$

$$\psi_T = \Delta\psi_T / (1+p)$$

$$v^E = 2\pi / G_E \sum_{j=1}^{N_T} r_j I^E_j$$

$$v^T = 2\pi / G_T \sum_{j=1}^{N_T} r_j I^T_j$$

where:

$\psi_P \, (r, z)$	flux on the plasma boundary due to the plasma current;
$\psi_E \, (r, z)$	flux on the plasma boundary due to external equilibrium currents;
$M^E_{ij} \, (M^T_{ij})$	mutual inductance between the j-th coil of the equilibrium (transformer) winding and the circumference passing at (r_i, z_i) on the plasma boundary;
$\varepsilon_E \, (\varepsilon_T)$	flux tolerance on plasma boundary for the equilibrium (transformer) winding;
N_B	number of controlled points on plasma boundary;
$N_E \, (N_T)$	number of possible equilibrium (transformer) coils;
$I^E_j \, (I^T_j)$	current in thj i-th equilibrium (transformer) coil;
$s^E_j \, (s^T_j)$	cross section of the j-th equilibrium (transformer) coil;
$G^E_j \, (G^T_j)$	current density in the j-th equilibrium (transformer) winding;
p	prepolarization coefficient for the transformer.

The quality index chosen corresponds to the total volume of the PF coils, which is usually directly related to the cost; this does not exclude the possibility of using other quality indexes (e.g. total amperturns).

An example of the way how the procedure works is shown in Fig.5. In Fig. 5a the permissible region for the location of PF coils is illustrated; Fig. 5b shows the PF currents at the beginning of the pulse (i..e. transformer winding); Fig.5c shows the external currents obtained for the plasma equilibrium.

2.4. OPERATING SCENARIO

The operating scenario (Fig. 6) consists of the following phases:
- startup: the plasma is formed in the chamber, the current is initiated, then ramped-up and finally the plasma is heated to ignition;
- burn: the plasma remains in thermonuclear condition releasing fusion power to the wall; the plasma current as well as the other parameters are constant;
- shutdown: the plasma is cooled and the current is ramped-down to zero;
- dwell: the machine is reset for the following cycle.

The duration of the burn phase depends on the method of plasma current generation. If a non-inductive method is used the transformer is no longer required and the burn phase is not upper bounded.

The advantages of having a steady state reactor are quite obvious but there are still uncertainties on its physical feasibility.

For this reason large tokamaks are usually designed assuming as reference a fully inductive method of plasma generation. In this case the duration of the burn phase is limited by the capability of the transformer to provide the ohmic flux required during this phase.

The dynamics of the plasma parameters during the cycle can be considered slow if compared with the typical time constants of the metallic structures. As a consequence the cycle can be reconstructed by means of a collection of static equilibria computed with the above illustrated procedure. The resulting PF coil currents are shown, in the NET case, in Fig. 7.

During startup the plasma equilibrium configuration varies considerably. As an example Fig. 8 shows how the plasma is grown to its final shape.

The intrinsic disadvantage of this approach is that the single equilibria are completely independent from each other. Therefore due to the non-linearity of the MHD equilibrium problem, it is not guaranteed that the plasma really evolves along the trajectory one obtaines connecting the single static equilibria.

A more self-consistent approach is possible making use of MHD evolutive codes where the plasma equilibrium is coupled to diffusion equations [13, 14].

2.5. FIELD PENETRATION

In the previous sections it has been shown that static models can be satisfactorly employed for the design of the PF system. However some time-varying effects, related to the field penetration, should be considered.

In fact the vacuum vessel shields the plasma from the penetration of the electric and magnetic fields during startup.

2.5.1. Electric Field Penetration. The plasma current is created by means of a suitable rate of

198

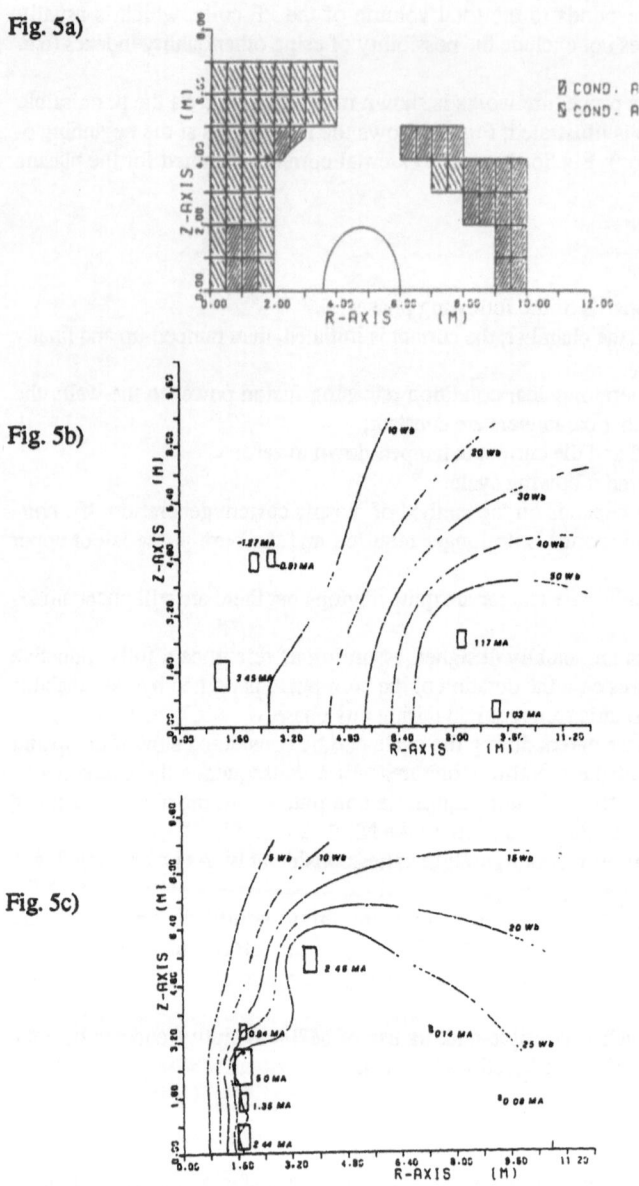

Fig. 5a)

Fig. 5b)

Fig. 5c)

Fig. 5   Example of PF coil design.

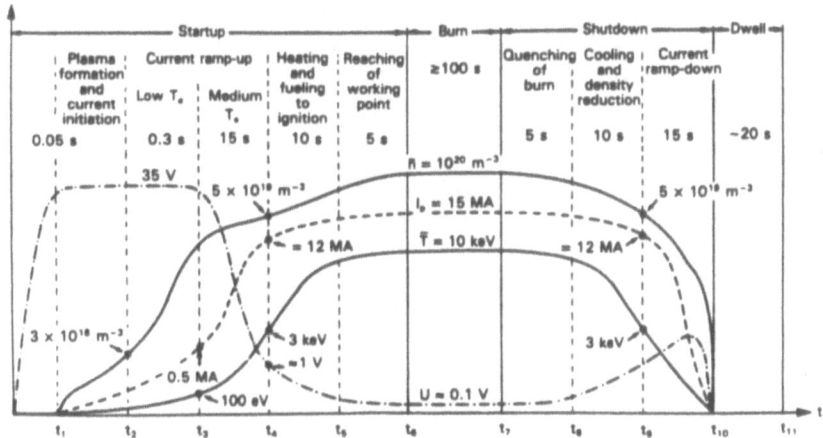

Fig. 6 Typical operating scenario in NET. '

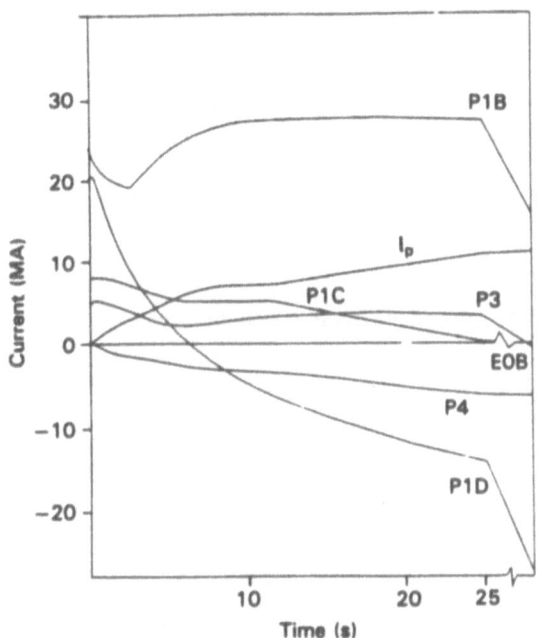

Fig. 7 PF coil current variation in NET during the cycle.

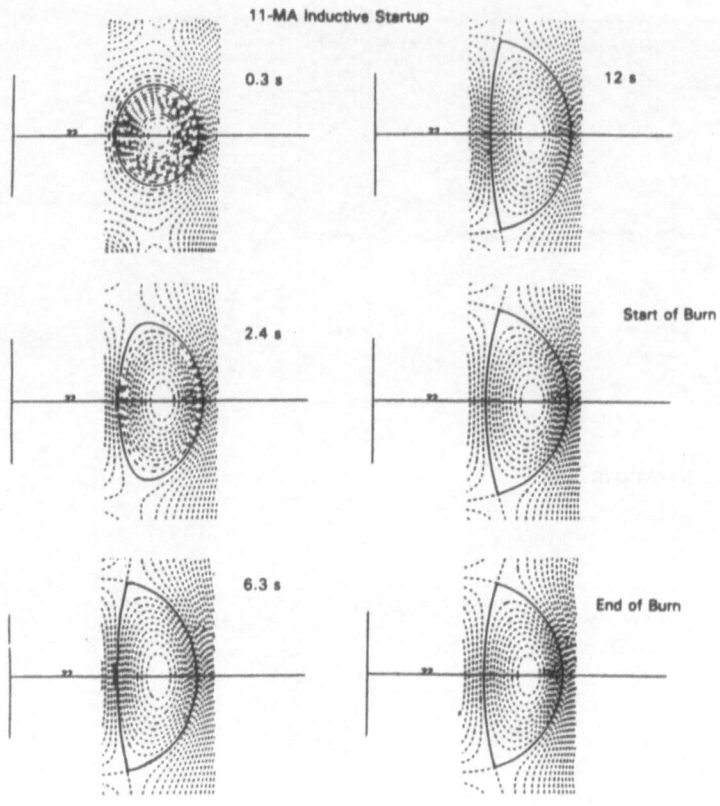

11-MA Inductive Startup

0.3 s

12 s

2.4 s

Start of Burn

6.3 s

End of Burn

Fig. 8 Evolution of the plasma configuration in NET troughout the cycle.

change of magnetic flux linked with the plasma ring. For a typical design of a tokamak reactor, the magnetic flux swing induces toroidal currents on both the plasma and the surrounding toroidal metallic structures. The loop voltage V_P induced in the plasma is then practically determined by the total resistance of the vacuum vessel, measured in toroidal direction. As an example Fig. 9 shows how, in INTOR [15],the required breakdown voltage is delayed by the presence of vacuum vessel.

From the viewpoint of the transformer design this effect implies for the plasma a loss of Vs decreasing with the toroidal resistance.

2.5.2. *Magnetic Field Penetration.* The magnetic field at the plasma results both from the external field (from PF coils) and fields of toroidally induced currents. Shielding properties of the vacuum vessel depend primarily on its toroidal resistance. Fig. 10 illustrates the effect of magnetic field screening in INTOR [15]. The magnetic field is the sum of external field B_{ext} (from PF coils) and eddy current field (from toroidally induced currents flowing in the vacuum vessel). To guarantee the plasma equilibrium since the initiation of the plasma current, the breakdown should occur when the total magnetic field in the plasma center is zero. For this reason in INTOR case the start of the transformer current swing is delayed of about 150 ms; in this way taking into account the penetration delay of the electric field (about 70 ms as from Fig. 9) the breakdown occurs when the total magnetic field is zero.

3. Control of the magnetic configuration

As a difference to present day devices, next generation tokamaks shall have long burn phases and complex startup procedures with significant changes of the plasma parameters throughout the operating cycle.

This makes the plasma position and current control one of the critical issues for the overall design of the machine, because of the constraints introduced on the reactor concept (e.g. maintenance scheme, passive structures, AC losses on superconductors, diagnostics, etc.).

The inherent multivariable and time-varying nature of the system suggests the use of modern control theory, provided that the modeling of the system is well suited to the application of a wide class of control techniques.

Therefore, in order to be able to guarantee the desired current, position and shape of the plasma, a feedback system has to be designed and analyzed.

The system includes plasma, conducting structural components (i.e. first wall, blanket, shield, vacuum vessel, stabilizing passive loops and others), PF and active control coils, power supplies, detectors, computers, etc.

For a typical next generation large tokamak, the scheme which is generally proposed is of the type shown in Fig. 11; it can be seen that the closed loop control of the plasma is performed by two separate systems:
i)   slow control system (second time-scale);
ii) fast control system (millisecond time-scale).

3.1 SLOW CONTROL SYSTEM

The slow system is primarily dedicated to the control of the radial position of the plasma. Since

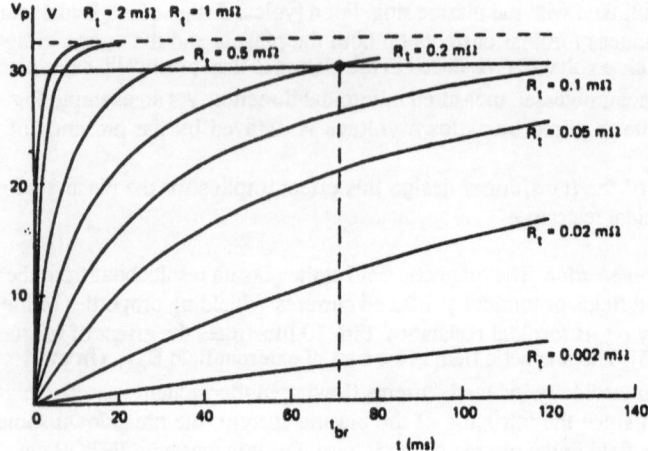

Fig. 9 Voltage induced in the plasma as function of time for various toroidal resistances in INTOR.

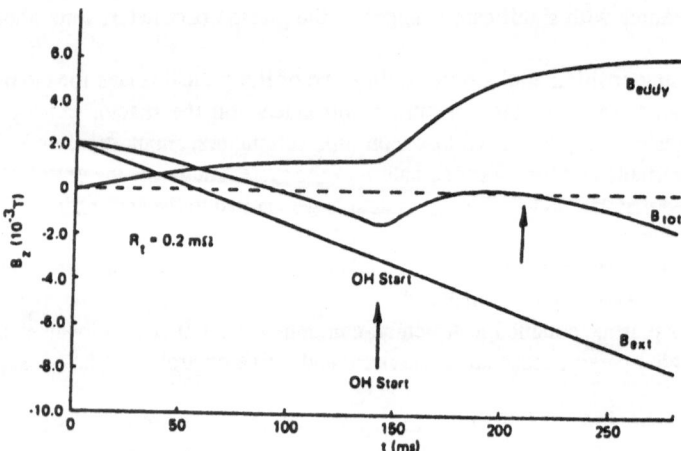

Fig. 10 Magnetic field at the plasma center during startup in INTOR.

the plasma ring is stable against radial displacements, the intervention time of the control is not a stringent requirement. On the other hand, the control field required to modify the radial position is quite large since it has to be compared with the existing vertical field produced by the PF coils; as a consequence, an intervention time on a very fast time scale would lead to an impracticably large power requirement.

The straightforward way of implementing this kind of control is to introduce, in the power supply scheme of the outer PF coils, an amplifier driven in feedback by a detector of the radial position of the plasma.

In a similar way, an amplifier introduced in the power supply scheme of the inner PF coils can be used for the control of the total plasma current.

When generalizing this concept, i.e. assuming that the power supply is such that the current in each individual PF coil can be changed independently from the others, it is possible to control even the shape of the plasma.

As a matter of the fact, it has already pointed out that the plasma shape will change substantially during startup, as will current, pressure and temperature profiles; profiles will continue to evolve during burn as accidental changes of the plasma parameters take place.

Having assumed that the shape control has to be performed by acting in feedback on the PF coil currents on a slow time scale, it is acceptable to formulate the problem of the shape control in terms of static MHD evolution.

The main task in the study will then be to formulate the MHD equilibrium equations, the circuit equations coupling plasma current, coil currents and eddy currents and transfer functions of power supply, computers and detectors and to develop a simulation code for the analysis of the time evolution of the total system.

In general, to simulate the plasma completely one should consider particle and heat transport, magnetic field diffusion, pressure balance, interaction with the wall, in a at least two-dimensional space.

In this case, the associated code would be too large and heavy to be used for both design and simulation purposes. For this reason, it is proposed to differentiate among the design models to be used for the evaluation of the main parameters of the control system and simulation models to be used to predict the behaviour of the plasma subject to given disturbances and control laws.

For the latter type of models, the already mentioned MHD evolutive codes [13, 14] can be conveniently used, provided that suitable pre- and post-processors are prepared for the manipulation of the selected control laws and global plasma parameters.

For the former type of models, a lumped parameter model of the PF system reveals to be particularly useful. For example, in one of the approaches proposed [16], the plasma is considered as a body having a discrete number of degrees of freedom corresponding to the shape parameters and acted on by the following forces:

a) forces due to electromagnetic interaction with external circuits;
b) forces due to eddy currents flowing in the passive structure;
c) forces due to externally applied magnetic fields;
d) additional forces of non-electromagnetic nature.

3.2 FAST CONTROL SYSTEM

The fast system is mainly dedicated to the control of the vertical position of the plasma. It is generally agreed [15] that an elongated plasma cross-section is advantageous in many respects for a tokamak reactor. Unfortunately, to maintain in equilibrium a vertically elongated plasma an external field with a negative curvature is needed [17], which makes it unstable against vertical displacements; therefore, the intervention time of the control must cope with the growth time of the instability.

Bearing in mind that the typical growth time is expected to be in the 10 ms range, the feedback control system has to act on a ms time scale. The radial control field required to counteract any accidental vertical displacement of the plasma can be provided by a pair of coils (Fig. 11) carrying the same current in opposite directions.

Most of the approaches to the design of the passive and active control system for the stabilization of an elongated plasma are based on a rigid displacement model for the plasma [15] [18]. This model can be regarded as a particular case of the already mentioned lumped parameter general model [16]. The main assumption is that the plasma is treated as a massless rigid conducting body carrying the toroidal plasma current; the body is free of moving vertically without changing the current density distribution; the model is then described by the following linearized equations:

$$L \dot{I} + R I + f \dot{z} = V \qquad (4)$$
$$F' z + g^T I = 0 \qquad (5)$$

where:

I is a vector containing the currents flowing in the stabilizing circuits, i.e. the metallic structure surrounding the plasma and any other passive or active coils ;

z is a scalar representing the vertical displacement of the plasma;

L and R are the inductance and resistance matrices of the stabilizing circuits;

f is the set of contributions to the fluxes linked with the stabilizing circuits, induced by a unit displacement of the plasma;

g is the set of the vertical forces acting on the plasma, produced by unit currents flowing in the stabilizing circuits;

V is the set of the externally applied voltages (zero for the passive circuits);

F' is the derivative of the destabilizing force with respect to z, i.e. the vertical force produced by the curvature of the equilibrium field on a plasma subjected to an unit vertical displacement.

When deriving z from Eq. (5) (which represents the plasma momentum balance) and substituting into Eqs. (4), the model assumes the very simple form :

$$L^* \dot{I} + R I = V \qquad (6)$$

where:

$$L^* = L - F'^{-1} f g^T \qquad (7)$$

Eq. (6) can be regarded as the circuit equation governing the stabilizing system, in which the inductance matrix has been modified to take into account the presence of the plasma.

Eq. (6) allows one to perform both the open-loop and the closed-loop analysis of the system.

The open-loop analysis aims to evaluate the stabilizing efficiency of the passive structure. In the absence of any passive structure, a plasma having an elongation as foreseen for the next generation tokamaks would be vertically unstable with a growth time in the microsecond range. Therefore a passive stabilizer is necessary to slow down the instability to values at which a feasible feedback controller is able to intervene.

The efficiency of the passive stabilization system is then measured by the growth time of the plasma in presence of surrounding non-ideally conducting metallic structures and in absence of active control; it can be easily computed when performing the eigenvalues analysis of Eq. (6). As a matter of the fact, for the cases of physical interest, the dynamic matrix of the open-loop system (i.e. the matrix $\mathbb{L}*^{-1}\mathbb{R}$) is a non-stability one: therefore the inverse of its largest (positive) eigenvalue is the searched growth time.

The open-loop analysis allows one to establish design guidelines for the plasma parameters and for the geometry of the passive structures.

The closed-loop analysis can be performed once the control law is specified. For example, for a simple PD controller, one has:

$$\tau \dot{V}_{active} + V_{active} = G(z + \tau_D \dot{z}) \tag{8}$$

where V_{active} is the voltage applied to the active circuit, τ is a suitable delay time constant (simulating the delays introduced by the power supply system), G and τ_D are the gains of the PD controller. The substitution of Eq. (8) into Eq. (6) allows one to perform the controller design, which consists in an optimal choice of G and τ_D.

Once the controller has been specified, Eq. (6) represents a tool to predict, within the assumptions of the model, the performance of the control system; a typical example is shown in Fig. 12. A better perfomance can be obtained if more advanced feedback control techniques are employed [19].

In order to make use of this agile design tool, the problem of the vertical stabilization has been extensively studied [15] using a simplified model for the plasma and a relatively rough schematization of the passive structures. However, the trend towards high plasma elongations has rendered the problem of the vertical stabilization one of the most critical issues in the design of a tokamak reactor. Therefore the necessity of limiting the severe constraints posed on the basic machine by the vertical instability calls for a more accurate analysis.

In this respect a physically more realistic model for the plasma displacement has been proposed [20], which, thanks to a method alternative to Eq. (7) for the computation the matrix L*, presents the advantage of being fully compatible with the above described design scheme, based on Eq. (6).

As far as the schematization of the passive structure is concerned, after an early phase in which only an axisymmetric modeling was adopted, correction terms to take into account the lack of axisymmetry of the structure were introduced [21] and studies with a 3D thin shell schematization were initiated [22]; a further improvement was more recently achieved [23], where a fully 3D eddy current code was suitably adapted for the computation of the terms $\mathbb{L}$, $\mathbb{R}$ and f appearing in Eq. (4).

The tendency of going towards a refined schematization of the passive structure presents the disadvantage of dealing with high order inductance and resistance matrices, which makes cumbersome the design of the feedback controller. For this reason a model reduction technique has been proposed [24], aiming to compact the required amount of information in a reduced order model with a moderate and controllable loss of precision.

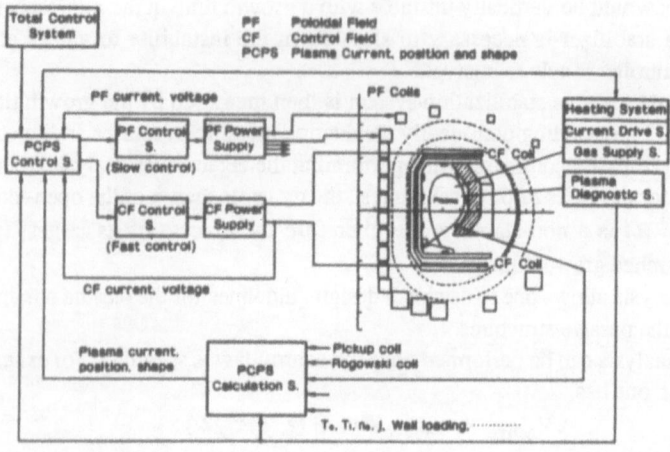

Fig. 11 Design concept for plasma current, position and shape (PCPS) control system in a t tokamak reactor.

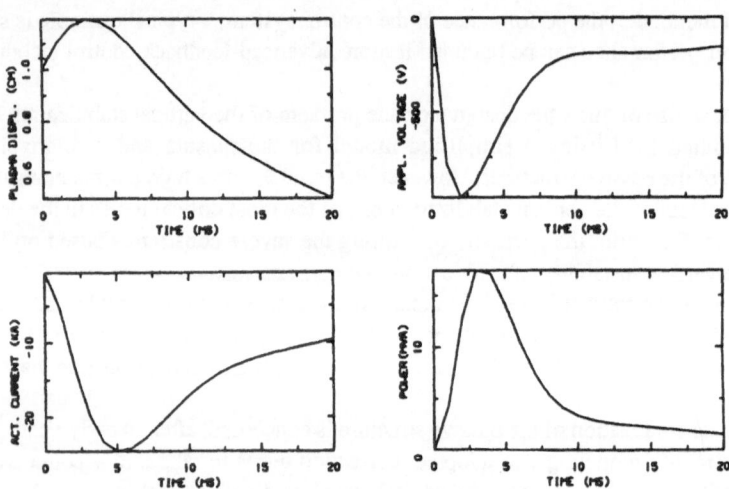

Fig. 12 Espected performance of an active feedback system in ITER after a sudden plasma displacement of 1 cm (I_P= 10.8 MA, growth time = 9ms, τ = 1ms, K = 50 kV/cm, τ_D = 10 ms).

4. Effects of plasma disruptions

In the present day tokamaks a non negligible number of discharges terminate abnormally in major plasma disruptions [25]; they are usually observed whenever some operational limits are violated, or in the case of failure of the feedback system for the control of the position of the plasma column in the vacuum chamber. Due to the lack of comprehension of the physical mechanisms occurring in a disruption and to the present practical impossibility to prevent them, it is generally agreed that also next generation tokamaks should be designed in such a way as to withstand their effects.

From the electromagnetic viewpoint one is mostly interested to the so called current quench phase of the disruption, during which the plasma currents vanishes rapidly. This phase is the consequence of the rapid falling down of the plasma temperature: the resulting highly resistive plasma terminates the current on a 10-100 ms time scale.

The quench of the plasma current has an heavy impact on the design of a fusion reactor, for the following two main physical reasons:

a) the magnetic energy stored in the plasma, which exceeds by far the thermal energy, disappears completely during this phase and impacts the surrounding structure: as an example, with a plasma current in the 20 MA range (typical value for next generation large tokamaks), the magnetic energy amounts to about 2 GJ;

b) due to the rapidity of the plasma current disappearance, the leading physical mechanism which is responsible for the absorption of the magnetic energy of the plasma, is the electromagnetic transfer of the current to the surrounding structure; as a consequence, due to the typical long range nature of the electromagnetic phenomena, all the metallic components of the machine are affected by the occurrence of a major plasma disruption.

When considering the typical components of a fusion reactor, the main electromagnetic effects of a plasma disruption can be summarized as follows:

- electromechanical loads: the interaction of the eddy currents with the already existing magnetic fields (poloidal and toroidal) gives rise to a body force distribution which affects considerably the design of the first wall and vacuum vessel;

- arcing: due to the lack of continuity of the first wall, during the early phase of the plasma current quench a loop voltage in the kV range is distributed across the small gaps of the structure, giving rise to possible arcing phenomena;

- a.c. losses on superconductors: the high dB/dt values are responsible for local heat deposition which could exceed the capabilities of the helium cooling;

- current changes on PF coils: the PF coil system has to withstand a total current change of the order of the plasma current on a time scale depending on the time constant of the vacuum vessel (50-100 ms typically).

It has to be remarked that the above effects depend strongly on the plasma behaviour during the disruption; for example, the possible motion of the plasma column during the quench changes qualitatively and quantitatively the pattern of the eddy currents.

A satisfactory modeling for the computation of these phenomena can be obtained by means of the quasi-stationary Maxwell equations, since the physical assumptions concerning the quench of the plasma current can be easily translated in terms of a source time-varying magnetic field; in this way purely electromagnetic computation tools, such as eddy current codes and associated pre- and post-processors can be conveniently employed.

One of the advantages of this approach is to separate the safety margins one has to introduce in the design because of the plasma disruptions into two categories:
- physical uncertainties on the plasma behaviour: these affect only the input data for the eddy current code;
- approximate solution of the electromagnetic problem: the error introduced depends on the features of the code and on the schematization of the structure.

With reference to the latter aspect, the approaches and associated numerical codes can be divided into three categories:
a) axisymmetric codes;
b) thin shell codes;
c) fully 3D codes.

The main advantage of the first type of codes, apart from their intrinsic geometric simplicity, is the possibility of a self-consistent linking with plasma evolutive equilibrium codes; therefore their use is advisable whenever the main interest is an accurate treatment of the driving time-varying field.

A much better schematization of the structure is possible with the second category of codes, where the geometry is represented by a number of thin shells and plates; their employment is well suited for the study of large scale effects, where the thickness of the structures is not expected to play a major role.

The obvious advantage of the last category of codes is that one has, in theory, the possibility of modeling the whole structure in all its details. However, due to practical computer limitations, a number of simplifications have to be introduced when schematizing the structure; their effect should be investigated [26] in order to gain confidence in the results achieved. Another advantage of this category of codes is the possibility of an easy interface with stress analysis codes, which are usually 3D in nature.

As an example of this methodology, we report here some of the results achieved for the vacuum vessel structure proposed for NET .

The objective of the NET project [27] is the design of a large magnetic confinement fusion device aimed at demonstrating the feasibility of energy production under the operating conditions of a reactor. The main components of the NET tokamak are shown in fig. 13.

The vacuum vessel (fig. 14) is a massive structure that constitutes the plasma chamber and support the plasma facing components.

From the electromagnetic point of view, the vessel can be schematized as a number of hollow massive parts (rigid sectors) connected through thin conductors of high resistivity. Taking advantage of the sector symmetry and the toroidal periodicity, it is possible to model only 1/32 of the whole structure. A possible mesh for a fully 3D eddy current analysis is shown in fig. 15.

The main results carried out assuming for the plasma disruption he scenario illustrated in Tab.1, are reported in Tab. 2.

The NET toroidal current I_{vv} in the vessel can be regarded as the sum of two contributions: the current flowing through the inner part (I_{vv1}) and that flowing through the portion of the vessel between the auxiliary heating ports and the vacuum ducts. In fig. 16 the NET currents I_{vv} is plotted with the contributions I_{vv1} and I_{vv2} .

Finally, Fig. 17 shows the current density distribution in the structure at the end of disruption.

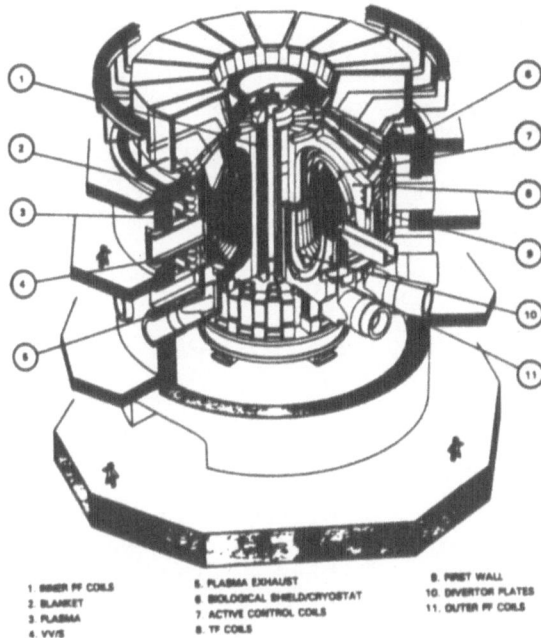

1. INNER PF COILS   5. PLASMA EXHAUST   9. FIRST WALL
2. BLANKET   6. BIOLOGICAL SHIELD/CRYOSTAT   10. DIVERTOR PLATES
3. PLASMA   7. ACTIVE CONTROL COILS   11. OUTER PF COILS
4. VV/S   8. TF COILS

Fig. 13   A view of the NET device.

1. BLANKET PORT   4. VACUUM PUMPS
2. HORIZONTAL ACCESS   5. PARALLEL SEGMENT
3. VACUUM DUCT   6. WEDGE SEGMENT

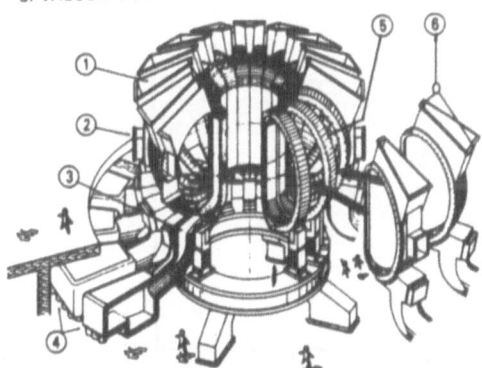

Fig. 14   The NET vacuum vessel.

Fig. 15   A 3D mesh of NET vacuum vessel.

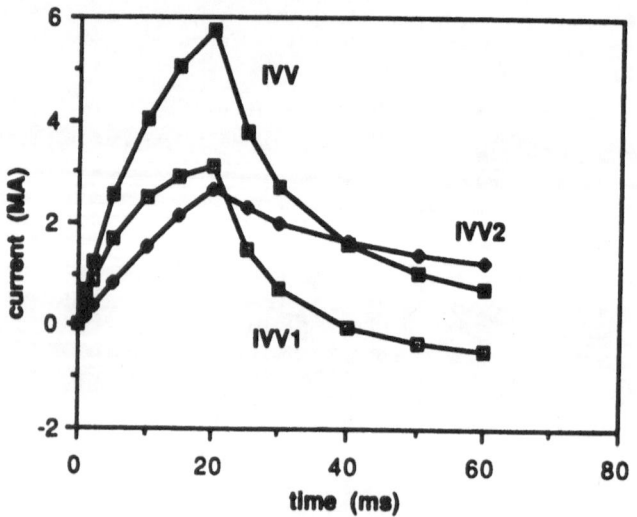

Fig. 16   The toroidal current in the NET vacuum vessel .

TABLE 1 . Plasma disruption scenario in NET

Plasma current	15 MA
Current quench	linear decay in 20 ms
Plasma current distribution	filamentary
Location of plasma current	R = 5.41m, Z = 0
Plasma motion	no
Toroidal field	4.8 T at R = 5.41m

PF. coils (end of burn)	R (m)	Z (m)	I (MA)
PF1B	1.48	5.00	3.77
PF1C	1.48	3.00	- 7.14
PF1D	1.48	1.25	-25.00
PF3	4.05	6.75	6.69
PF4	9.65	2.50	- 7.84
connections	up-down series connected and short circuited (zero resistivity)		

Passive structures		
vacuum vessel	16	modules
torus resistance	0.2	mΩ
first wall/blanket	48	modules

TABLE 2 . Eddy currents in NET due to the plasma disruption

Toroidal current in the vacuum vessel at $t=\tau_d$=20 ms (end of plasma current quench):

I_v	(total current)	5.77	MA
I_{vv1}	(inner contribution)	3.11	MA
I_{vv2}	(outer contribution)	2.66	MA

Current changes in the PF coils at the end of the transient phase ($t>>\tau_d$):

2 x ΔI PF1B	1.57	MA
2 x ΔI PF1C	2.21	MA
2 x ΔI PF1D	7.24	MA
2 x ΔI PF3	1.35	MA
2 x ΔI PF4	2.87	MA

Ohmic power at $t = \tau_d$ in 1/16 of the structure: 696 MW

5. Conclusions

It has been shown how a number of electromagnetic problems arising in the design of a fusion reactor can be tackled with the help of suitable numerical procedures. It goes without saying it that an improvement and refinement of the available methodologies is generally needed; any further discussion on this point is not made here because it would require entering in detail on the features of the specific codes.

We limit instead ourselves on commenting on some topics that, at the Authors' opinion, should deserve more attention in the future. Without pretending to be exhaustive, we mention here:

- magnetic measurements: due to the thermonuclear character of the plasma in a fusion reactor, the design of the magnetic flux and field probes to be used for the detection of the position and shape of the plasma is a critical issue; typical problems are the optimal selection of their number and location and the eddy current effects on the measurements;
- startup strategy: the foreseen plasma has to be safely driven to the burn phase taking the best advantage from the power supply available; an open-loop optimization of the trajectory is needed;
- startup control: the plasma can behave differently from what planned for that specific pulse; an on-line fast identification of the plasma parameters has to be connected with the feedback control of the PF coils;
- zooming: when going towards the detailed design of the metallic structure of the reactor one should consider electromagnetic effects even on very small components; reliable ways are then needed to pass from the whole structure to a local detail;
- coupled problems: due to the large dimension of the reactor components electromechanical forces cause appreciable elastic displacements, which could give rise to an interaction with other field components.

Another general comment arising from this review of electromagnetic problems in fusion reactor design is the necessity of an integration between codes of different nature. A chain of codes is for example realized whenever the data needed to run a specific code are taken from the output of another code; obviously, suitable interface routines have to be prepared in this case. In this simple way of integrating codes, the level of accuracy degrades when advancing along the chain. Taking instead advantage of modeling overlaps, it is sometimes possible to realize closed-loops among different codes [28]; the advantage of this procedure is the reduction of schematization errors and a general gain of confidence on the results achieved.

May we conclude the present note with a phylosophic statement. The development and use of computer codes of increasing sophistication is to be encouraged, since powerful tools are needed to tackle the formidable complexity of the design problems one encounters in a fusion reactor; but, at the same time, they should never be regarded as a substitute of human skill or expertise.

REFERENCES

[1]   Wesson, J. (1987) Tokamaks, Clarendon Press, Oxford.
[2]   Pasotti G. and Ricci M.V. (1986) Final Report on electromagnetic effects of the use of

ferromagnetic materials within the NET Machine', Frascati (Italy).

[3] Schmidt, G. (1966) Physics of High Temperature Plasmas, Academic Press, New York.

[4] Krall, N.A. and Trivelpiece, A.W. (1973) Principles of Plasmas Physics, Mc Graw-Hill, New York.

[5] Grad, H. and Rubin, H. (1958) 'Hydromagnetic Equilibria and Force-free Fields', Proc. of 2nd U.N. Conf. on the Peaceful Uses of Atomic Energy, Geneva, Vol.31, 190.

[6] Shafranov V. D. (1958) 'On Magnetohydrodynamical Equilibrium Configurations' Soviet Phys. JETP 6, 545-554

[7] Maschke E.K. (1973) 'Exact solutions of the MHD equilibrium equations for a toroidal plasma', Plasma Physics 15, 535-542.

[8] Goedbloed, J. P. (1982) 'Free-boundary High-beta Tokamaks. I. Free-boundary equilibrium', Phys. Fluids 25, 852-868.

[9] Feneberg, W. and Lackner, K. (1973) 'Multipole Tokamak Equilibria' Nuclear Fusion 13, 549-556.

[10] Lackner, K. (1976) 'Computation of Ideal MHD Equilibria' Computer Physics Comm. 12, 33-44.

[11] Salpietro E. et Al. (1988) 'Next European Torus Operation Cycle', Fus. Tech. 14, 145-155.

[12] Wilhelm R. (1989) 'Plasma Heating. A Comparative Overview for Future Applications', Proc. of the 15th SOFT ,Utrecht 1988, North-Holland, 167-180.

[13] Jardin S.C., Pomphrey N. and De Lucia J. (1986) 'Dynamic Modeling of Transport and Positional Control of Tokamaks' J. Comput. Phys. 66, 481.

[14] Albanese R., Blum J. and De Barbieri O. 'Numerical Studies of the Next European Torus via the PROTEUS code', Proc. of the 12th Conf. on Numerical Simulation of Plasmas, San Francisco, 1987.

[15] INTOR GROUP (1986) International Tokamak Reactor, Phase Two A, Part II (Rep. Int. Workshop Vienna, 1984 e 1985), IAEA, Vienna ,1986.

[16] Ambrosino G., Celentano G., Coccorese E. and Garofalo F. (1989) 'Controller Design for Plasma Position and Current Control in NET', Proc. of the 15th SOFT ,Utrecht 1988, North-Holland, 1670-1674.

[17] Muskhvatov, U.S. and Shafranov U.D. (1971) 'Plasma Equilibrium in a Tokamak' Nuclear Fusion 11, 605-633.

[18] Coccorese, E. and Garofalo F. (1985) 'Plasma Position Control', in Knoepfel (ed.), Tokamak Startup,Plenum Press, 337-349.

[19] Albanese R., Ambrosino G., Coccorese, E. Garofalo F. and Rubinacci G. (1989) ' An alternative Approach to the Analysis of the Active Vertical Stabilization in a Tokamak', Proc. of the 16th European Conference on Controlled Fusion and Plasma Physics, Venezia 1989, 447-450.

[20] Albanese R., Coccorese E. and Rubinacci G. (1989) ' Plasma Modeling for Vertical Instabilities' Nuclear Fusion 29, 1013-1023.

[21] Bobbio S., Coccorese E., Fabricatore G., Martone R. and Rubinacci G. (1985) 'Passive Control of the Vertical Instability in INTOR', Fusion Technology 7, 345-360.

[22] Nakamura, Y and Ozeki T. (1983) 'Eddy Current Analysis in JT-60', Proc. of the 12th SOFT ,Julich 1982, Pergamon Press, 359-346.

[23] Albanese R., Coccorese E., Martone R. and Rubinacci G. (1989) 'Analysis of the Vertical Instabilities in Air Core Tokamaks in the presence of Three Dimensional Conducting Structures', COMPUMAG '89, Tokyo.

[24] Coccorese E., Coco S. and Martone R. (1989) 'A Model Reduction Technique for High

Order Eddy Current Problems', COMPUMAG '89, Tokyo.

[25] Wesson J.A. et Al. (1989) 'Disruptions in JET' Nuclear Fusion 29, 641-666.

[26] Coccorese E. and Martone R. (1989) 'Electromagnetic Computations on Fusion Devices: Various Structure Schematizations in the Next European Torus', Fusion Technology 16, 514-520.

[27] Toschi R. et Al. (1988) 'Next European Torus Objectives, General Requirements, and Parameters Choices', Fus. Tech. 14, 145-155.

[28] Albanese R., Coccorese E. and Rubinacci G. (1988) "Computation Methods for Transient Electromagnetic Studies in Tokamak Devices", Proc. of Beijing International Symposium, Pechino 1988, Pergamon Press.

ELECTROMAGNETIC ANALYSIS APPLIED TO THE TOKAMAK FIRST WALL DESIGN

Y.R. CRUTZEN
Commission of the European Communities
Joint Research Centre (JRC)
Institute for Systems Engineering and Informatics (ISEI)
21020 Ispra (Va) – Italy

ABSTRACT. In a Tokamak-type fusion device, the presence of large magnetic fields and the plasma disruptive instabilities generate electromagnetic-type phenomena in transient conditions which lead to strong induced eddy currents and body loads. These electromagnetic loads must be correctly accommodated in the engineering design of all the reactor components. In particular, the plasma disruptions have a significant impact on the mechanical design of the internal reactor components, such as the multi-segmented first wall (FW), and of their supports. Fully three-dimensional (3-D) dynamic analyses are required due to the geometry complexity of the "conducting" and "plasma" regions and to the transient nature of the phenomena, both from the electrical and mechanical standpoints.

1. INTRODUCTION

Three-dimensional (3-D) computer simulations occupy an increasing place in analysing time-varying structural problems. The use of digital computers and graphic equipments and the development of recent numerical methodologies make them today the most promising line of research. In particular, there is an actual need to investigate 3-D magnetodynamic problems, even if they require complex mathematical formulations.

An effort has been made to improve the computerized 3-D design/analysis capabilities at the ISEI Institute of the JRC and, particularly in the context of electromagnetic computations, with the collaboration of Universities and European leader Companies, to satisfy the main requirements for the treatment of 3-D time-dependent engineering problems. Recent 3-D software tools, pre- and post-processing facilities and interface capabilities between CAD/FEM/CG technologies have been developed and applied.

In the context of protection of next magnetic fusion reactor systems against plasma disruption damages, numerical simulations of the electromagnetic/mechanical interactions on Tokamak in-vessel components are presented. Different computer codes are applied for the

215

Y. R. Crutzen et al. (eds.), Industrial Application of Electromagnetic Computer Codes, 215–231.
© 1990 *ECSC, EEC, EAEC, Brussels and Luxembourg.*

resolution of time-varying eddy currents induced on 3-D non-axialsym-
metrical but segmented passive conductors. The investigations into the
electromagnetic analysis of major plasma disruption scenarios (fast
current time-decay and rapid vertical movement) and their mechanical
effects are considering the complex 3-D geometry and the varying electrical resistivities of the plasma-facing components (i.e. first wall
(FW), shielding panel, divertor plate and vacuum vessel (VV)).

In relation with the mechanical support choice, some design issues for the fastening and guiding systems of the removable FW segments are proposed.

In addition, the presentation of computerized animation movies,
such as time-histories of eddy current circulation and related structural deformation behaviour, enhances the understanding of such 3-D
complex modelling and numerical analysis.

2. THE TOKAMAK DEVICE AND THE PLASMA DISRUPTION EVENTS

The Tokamak (Toroidal Magnetic Chamber in Russian) is one of the most
promising magnetic confinement systems to create the conditions for
nuclear fusion. The plasma is an ionized gaseous fuel made of charged
particles and heated to very high temperatures (100 million degrees
centigrade) in order to overcome the mutual electrostatic repulsion
between the nuclei involved in the thermonuclear fusion process. Among
the possible fusion reactions involving light nuclei, the one of
greatest interest is that between Deuterium (D) and Tritium (T), the
two heavy isotopes of Hydrogen, which produces the element Helium
(^{4}He) plus a neutron (n), and which is combined with a large release
of energy (E) (see Casini /1/).

Of course, it is expected that the energy E produced from the fusion reaction exceeds the amount of energy required to run the system
and to heat the plasma (by the circulation of very large currents, the
injection of energetic atoms or the use of radiofrequency waves). In
addition, the hot plasma must be confined in isolation (inside a non-
material container), far from the vacuum chamber (VV) and the FW components, avoiding contamination and cooling, for a long enough period
of time (Lawson criterion: density x confinement time greater than a
few times 10^{20} s.m^{-3}). In the Universe, fusion plasmas (the sun and
stars) are held together by large gravitational forces. In a Tokamak
machine, the plasma must be confined and driven by suitable magnetic
field configurations: PF and TF coils are distributed all around the
machine (see Bobbio in /1/).

Up to now, a few large experimental Tokamak machines have been
built all around the world (e.g. JET, TFTR, JT-60, T-15, TORE-JUPRA,
FTU, ASDEX, etc.) (see Bertolini in /1/). A particular but characteristic is that all of them exhibit plasma disruptive instabilities.
Among the observed plasma disruptions, two distinct scenarios can be
actually identified:
- "disruptive pulse" phenomena occurring during normal operation when
  a critical plasma density is reached: high energy losses and increased impurity radiation are associated with the formation of un-

stable magnetic islands that destroy the confinement;
- "vertical instability" events occurring when the plasma position growth rate exceeds the capability of the feedback stabilisation system of the machine.

For the JET (Joint European Torus) (Rebut et al. /1,2/) reactor, systematic studies are dedicated to a better knowledge of the plasma behaviour during a disruption event (rapid energy dissipation and quenching of plasma current). Even if the plasma disruptive instabilities will be better controlled and their frequency reduced in the future, the performance of the plasma vertical stability is related to a failure condition of the machine equipment and has to be considered in any case.

As a consequence, a serious engineering issue in fusion reactor design is the complete knowledge of the effects of the electromagnetic-type phenomena which occur during any plasma disruption scenario. This accidental event creates a drastic release of a large amount of energy in the passive structures surrounding the plasma ring (FW and VV) and could cause material damage, support systems loss and structural instability problems.

The effects of reference plasma disruption scenarios on the structural integrity of the FW, foreseen for next generation Tokamaks such as NET (Next European Torus) (Toschi et al. /1,3/) and ITER (International Thermonuclear Experimental Reactor) (Tomabechi et al. /4/), are investigated in the next paragraphs.

3. 3-D EDDY CURRENT ANALYSIS: MODELLING EFFORTS AND COMPUTER TOOLS

3.1 Modelling Requirements

As far as the fusion engineering studies are concerned, several modelling efforts have been performed in order to account for and to correctly simulate the following situations (Turner /11,13,16/, Salpietro et al. /10,13/) (see also Figs.1 and 2):
- the plasma instability nature and scenarios characterized by fast quench of a large current and coupled with a rapid plasma movement;
- the complex magnetic field distribution inside the machine in order to drive and to confine the plasma column;
- the 3-D in-vessel multi-segmented conducting bodies with narrow gaps between them allowing remote maintenance (see Fig.3);
- the inductive coupling between the FW segments, the VV and coils and the electromagnetic interaction between the "plasma" and "conducting" regions;
- the effects of the electrical breaks or connectors which modify the circulation of eddy currents and of the active coils and saddle loops included for the plasma position control.

3.2 Computer Tools, Formulation and Numerical Techniques

During the last few years, a few computer codes have been developed for the resolution of transient eddy currents induced in 3-D passive

non-magnetic conductors. In the context of the present modelling re-
quirements, the following computer tools have been available and ex-
tensively applied during the last period of time:
- UNISH and SCILLA codes developed in collaboration with Naples and
  Salerno Universities and installed at JRC-Ispra (Rubinacci /5/,
  Crutzen and Rubinacci /15/);
- CORFOU code coming from CEA (Comité de l'Energie Atomique, Saclay)
  and available at NET Team (Garching) (Blum et al. /6/);
- CARIDDI 3-D code developed at Naples and Salerno Universities
  (Albanese and Rubinacci /7,9,13/);
- TRIFOU 3-D code belonging to EdF/DER (Electricité de France,
  Clamart) (Bossavit, Verite /8,9,14/) and applied for JRC study con-
  tracts.
     The codes are based on different formulations for computing the
eddy currents from the quasi-static form of the Maxwell's field equations and they consider different unknowns for 3-D eddy current computations (Trowbridge et al. /13,16/, Albanese and Rubinacci /19/):
- the magnetic field potential A (combined with the electric scalar
  potential V);
- the electric vector potential T (coupled with the magnetic scalar
  potential ϕ);
- the magnetic field vector H (also combined in the scalar region with
  the potential ϕ).
     The 3-D domain can be divided into a "source" region (i.e. asso-
ciated to the plasma ring and the external coils in magnetic fusion
devices), a 3-D shaped "electrically conducting" media and an
"air/void" region all around. The open boundary problem leads to various numerical modelling approaches based on the finite element
methodology (FEM) (Albanese et al. /9,13,16,19/, Bossavit et al.
/9,14,16/):
- FEM integro-differential formulations;
- mixed FEM-boundary integral element method (BIEM).
     The codes UNISH and SCILLA have been conceived as special-purpose
tools for computing eddy currents and resulting transient electromagnetic forces induced in fusion reactor systems. They are applied to
non-magnetic conducting thin shells, i.e. that their thickness is
small compared to the skin depth of the phenomena, and their
formulation is based on the A-V method. The numerical treatment is
based on the discretization of the conductor into a mesh of 4-node
isoparametric shell elements, assuming the current density uniformly
distributed along the thickness.
     The code CORFOU computes the eddy currents in structures which
can be assimilated to thin shells. The code is based on the T-ϕ
formulation of the eddy current problem. Using the finite element
formulation, the conducting surface is discretized into triangles.
     The code CARIDDI is an advanced computer tool dedicated to fusion
reactor applications. It computes the eddy currents induced in 3-D
passive non-magnetic conductors considering a T-ϕ approach. In
CARIDDI, 8-node isoparametric brick elements are implemented using the
edge variable element. As a matter of fact, CARIDDI allows to analyse
induced eddy currents in both massive structures and shells, which

typically coexist in recent high energy devices.

The 3-D code TRIFOU ("Calcul TRIdimensionnel de Courants de FOUCAULT") is a more general-purpose package for magneto-static and magneto-dynamic problems, based on the H- formulation. The use of edge variable elements enforces the correct interface conditions. Concerning the 3-D numerical modelling, a macro-mesh of the conductor is first created by means of a reduced number of brick-type finite elements. Then, the complete mesh of tetrahedra elements is automatically generated.

As far as their mathematical formulations and the specific numerical techniques are concerned, more detailed aspects are outlined by the previous referred authors who are also lecturers of the present EUROCOURSE.

3.3 Modelling Approaches

From the analyst's standpoint, the FEM-based computer programs can be subdivided into two distinct modelling approaches (Crutzen et al. /13,14/):
- "shell" model considering a surface conductor where only two components of current ("2cc") in each finite element are computed;
- "solid" model modelling the thick conductor with three current components ("3 cc") in each elementary volume.

In the presence of 3-D multi-segmented passive conductors where all parts are inductively coupled, using a network approach, the efficiency of these formulations is closely connected to the calculation of the global inductance matrix which requires careful optimization because it is by far the most computationally expensive (up to 80% of the CPU time). The integration in time of the differential equations is often based on the Crank-Nicholson scheme /9,13,14,16/.

In practice, the "2cc" codes (UNISH, SCILLA, CORFOU) are faster and simpler to use than completely 3-D codes, since the computation of the inductance matrix requires only one surface integration and since the modelling refers only to equivalent shell surfaces. For these reasons they are useful particularly in the predesign phase, when many different geometries must be analysed and compared (see next paragraph).

Of course, fully 3-D thick conductors should be analysed by other programs, "3 cc" codes (such as CARIDDI and TRIFOU), where the inductance matrix computation requires complete volume integration /9,13,14,15/. In addition, they incorporate "edge elements" which are superposed to the hexahedra or tetrahedra finite elements. In general, such codes are used for detailed studies and are considered as essential tools for the final design phase.

All the computer programs at our disposal calculate first the induced eddy current circulation in space and time in the structural component and then determine the electromagnetic body force distribution arising from the interaction with the large magnetic fields. The finite element modelling incorporates the active part with all the external fields, plasma region and axisymmetric conductors, and the passive part which represents the electrically conducting components carrying the saddle currents.

The plasma ring can be discretized considering the minor and major axes of its elliptical shape and including a current density profile. Otherwise, the plasma column can be modelled by a set of axisymmetric (toroidal) rectangular or circular filaments located at the plasma major radius. This is also useful to simulate the vertical motion (vertical instability).

The definition of the boundary conditions is sometimes difficult and complex, but essential in order to define the problem correctly and to exploit the symmetry conditions (reflection and rotation). In the present applications, the segmented in-vessel components are facing the plasma column around the revolution axis of the Torus.

An additional critical issue is related to the correct simulation of a very narrow gap (about 2 cm) between each disconnected metallic box or panel (see Fig.3 and next paragraph).

As far as the examples of application of the computer programs are concerned, the following different parts are outlined:
- electromagnetic/mechanical analysis of NET FW system;
- plasma disruption analysis on the ITER FW system.

4. ELECTROMAGNETIC-MECHANICAL ANALYSIS OF NET FIRST WALL SYSTEM

The first study refers to the NET I design of the double-null plasma configurations which are foreseen during the initial physics phase of the machine operation for the inboard FW and during the final technological phase when a breeding blanket will be installed in the outboard region /1,3/. The difficulties for the correct simulation of the induced currents are that, in contrast with existing devices (JET, TORE-SUPRA, TFTR, JT-60) where plasmas are only surrounding thin axisymmetrical vacuum chambers (VV), 3-D segmented in-vessel components allowing remote maintenance are designed for NET (see Farfaletti-Casali in /1,3/).

Previous studies (Crutzen et al. in /12/) have already illustrated how the high segmentation of the internal structural components reduces the amplitude of the induced loads during electromagnetic transients. Both inboard and outboard regions are consequently split into 48 internal removable segments allowing remote maintenance (see Fig.1). Each inboard module corresponds to elongated thin shielding panels, water-cooled, while each outboard module consists of a closed and tight box-like thin structure enclosing the breeding blanket units (not considered in the calculations) (see Fig.2). According to the NET FW concept, the main plasma-facing walls are relatively thin structures in austenitic stainless steel including poloidal cooling tubes.

As far as the outboard FW segments are concerned, a first reference plasma disruption scenario (current quench of 10.8 MA/20 ms) has been simulated. As shown in Fig.4, the application of the different computer codes has permitted to perform cross-examinations which can

be summarized as follows /13,14,16/:
- similar eddy current circulation in closed loops with maximum amplitude (15% difference) and with delay in penetration (see Fig.5);
- discrepancies in eddy currents localized in the stiff back region
  where a "3 cc" model can encounter real 3-D effects only (see
  Fig.6);
- eddy current gradients through the thickness are simulated accurately by "3 cc" codes only.

The comparison is extended to the complex evaluation of the
Laplace transient forces and torques acting on the FW component. Good
agreement is obtained for the force resultants (in x,y,z directions)
and for the electrical time constant (ratio between the stored magnetic energy and the dissipated ohmic power in 1/4 model), as shown in
Table 1.

TABLE 1: NET outboard FW segment (48 modules), plasma disruption
(10.8 MA/20 ms)

Electromagnetic force resultants
Ohmic power/magnetic energy/time constant

1/4 model	FX (MN)	FY (MN)	FZ (MN)	P (MW)	W (kJ)	(2W/P) (ms)
Shell (2 cc)	1.62	0.63	0.21	3.35	4.99	2.98
Solid (3 cc)	1.90	0.64	0.24	4.06	5.36	2.64

The large induced radial resulting forces FX are mostly responsible for the generation of twisting deformation and overturning
situation that must be balanced by appropriate mechanical boundary
conditions.

The uncoupled mechanical computations are performed considering
the same finite element mesh for a correct transfer of the body forces
distribution in space and time. This is a main drawback even if the
mechanical response of the FW structural component reflects the fast-
transient nature of the electromagnetic phenomena /12,15/.

In relation with the mechanical support choice, some design solutions for the fastening and guiding systems of the removable FW segments have been proposed considering the wide range of variations of
the maximum local reaction forces (from 0.4 to 1.07 MN) transmitted to
the vacuum vessel, combined with the maximum Von Mises stresses generated inside the plasma-facing wall (from 110 to 190 MPa) and with
the peak strain energy indicating the global structural stability
level (from 101 to 332 kJ) /12/ (see Fig.7). Complementary studies related to FW instability problems are presented in /18/.

An engineering animation/navigation process (generation of motion
pictures) during the electromagnetic/mechanical transient improves the
understanding of 3-D modelling/analysis (see Fig.8).

Considering the coupling between plasma motion and current
quench, a second plasma disruption scenario has been simulated: an ex-

ponential vertical plasma movement before the plasma current quench
/14,16/. During the vertical displacement of the plasma ring, a completely different eddy current circulation is generated: the induced
currents flow vertically from bottom to top (and top to bottom) of the
elongated box and tend to re-establish the equilibrium. Even if the
consequences of the current quench are still similar to the previous
conclusions, the additional vertical movement can disturb the metallic
component integrity (initial impulsive-type excitation of opposite
sign), as shown in Fig.9.

Experimental results from JET and information from validation
studies need to be investigated more in detail (see Noll in /11/).
Additional investigations are carried on during the ITER Specialists
Meetings in order to define some realistic reference "dynamic"
disruption scenario.

As far as the inboard FW shielding panels are concerned, the distribution of the eddy current densities and of the electromagnetic
body forces within the inboard panel (1/4 model in Fig.10), subjected
to a plasma current quench of 15 MA in 20 ms, is successfully compared. But, with respect to the outboard FW segments characterized by
vertically distributed eddy current loops, FW models are also influenced by eddy currents flowing horizontally inside their thickness.
Therefore, a more sophisticated solid (3 cc) model including a finite
element discretization through the panel thickness has been considered. As illustrated in Table 2, the force resultant FZ obtained by
such a realistic solid model is different from the values computed by
the previous approximate shell models. This example is a typical case
where a full solid (3 cc) model is requested to study a relatively
thin panel when the thickness is no more small compared to the skin
depth.

TABLE 2: NET inboard FW segment (48 modules), plasma disruption
(15 MA/20 ms)

Electromagnetic force resultants
Ohmic power/magnetic energy/time constant

1/4 model	FX (MN)	FY (MN)	FZ (MN)	P (MW)	W (kJ)	(2W/P) (ms)
Shell (2 cc)	1.42	0.20	0.164	0.68	2.50	7.4
Solid (3 cc)	1.68	0.26	0.683	1.82	2.80	3.0

5. PLASMA DISRUPTION EFFECTS ON THE ITER INBOARD FW SEGMENTS

The international fusion project, ITER /4/ (Fig.11), involves the design of a large experimental Tokamak machine, where a toroidal cur-
rent-carrying plasma is still driven and confined. Accidental events
that must be considered during the long-pulse controlled-burn are
still the plasma disruptive instabilities. With respect to the NET de-

sign, complexities are arising from the extended 3-D geometry of the inboard region (the elongated FW panels have a box shape), from the narrow gaps between segments (see Fig.3) and from the increased plasma current and magnetic field magnitudes (about 20 MA and 5 T).

A complete 3-D electromagnetic analysis of the inboard first wall of the ITER device has been performed. The use of the computer programs CARIDDI and TRIFOU has demonstrated that the estimation of the Foucault eddy currents and Laplace body forces in complex 3-D situations can be predicted accurately (Crutzen et al. /16/).

Concerning the eddy current situation arising from the strong electromagnetic transient (about 20 MA/20 ms plasma disruption), the solid approach shows different eddy current circulations in the plasma-facing front part, side walls and thick back plate than the one discussed earlier for the NET device. The Foucault eddy current circulates in the plasma-facing front part, side walls and thick back plate. Due to the gap between each FW module, no toroidal continuity is allowed but only closed loops are created. Fig.12 illustrates the horizontal components of induced currents which are circulating around the front/side/back parts in the equatorial region. We can observe the 3-D nature of the eddy currents through the wall thicknesses and the real physical delay between their generation in the back region with respect to the front part.

The fast-transient electromagnetic body load distribution (Laplace forces) acting on the structural component and generated from the eddy currents magnetic fields coupling are also determined. Their distribution is closely correlated with the eddy current circulations.

In particular, from the TRIFOU computations and taking into account the different steady-state magnetic fields (toroidal and poloidal) and time-dependent fields (generated by plasma and induced currents), the time evolution of the radial and vertical Laplace forces in each specific FW domain (front/side/back/upper parts) and their resultant are automatically generated. The main results are the following:
- electromagnetic forces of opposite sign are generated in the front and back parts on the box-type structural component;
- higher radial and vertical forces are induced in the side walls;
- maximum radial force reaches + 2.2 MN (in each 1/4 model);
- maximum vertical force reaches 1.4 MN (in each 1/4 model).

The use of "static" representations of a time-varying phenomenon (by "flashes" of information) is very often unsatisfactory for real 3-D time-varying computer simulations. Using advanced computer graphics techniques, an animation movie has been realized. In the present situation, the eddy current and electromagnetic force distributions are visualized by coloured arrows representing a vectorial variable moving/changing in space and time. A particularly clear picture of the complex manifestation of eddy currents and electromagnetic forces helps us in the interpretation of the consequences from an electromagnetic-mechanical point of view. Complementary studies related to the internal segmentation (by electrical insulation of each cell) are presented in Ref./18/.

6. CONCLUSIONS

Computerized analyses of the impact of electromagnetic-type phenomena caused by plasma disruption events on the in-vessel components of NET and ITER Tokamak fusion devices have been described. Shell (2 cc) and solid (3 cc) approaches have been extensively applied for cross-examinations.

In addition, the possible presentation of animation movies including time-dependent electromagnetic (eddy currents) and mechanical (transient and vibration) behaviour simulations are well recommended in order to enhance the understanding of such complex 3-D modelling and electromagnetic computations.

In the context of structural design and safety analysis of fusion devices, there is an urgent need to integrate validation studies in the presence of transient electromagnetic phenomena including both the electromagnetic and the mechanical aspects. The recent activities of our JRC Institute in that context are reported in Ref./17/.

ACKNOWLEDGEMENTS

The author wishes to thank Messrs. R.W. Witty, G. Volta, G. Casini, F. Farfaletti-Casali and M. Biggio for their interest in this research activity. The collaboration of colleagues from our JRC Institute, from the Electrical Engineering Institute of the Universities of Naples, Salerno and Genoa, from the NET Team in Garching and from the Research and Development Division of EdF (Electricité de France) is gratefully acknowledged.

REFERENCES

1. Ispra Courses, edited by G. Casini on 'Engineering Aspects of Thermonuclear Fusion Reactors' (1980) CEC-JRC, Ispra.
2. JET Joint Undertaking (1986/87/88/89) Progress Reports, ESCS/EEC/EURATOM, EUR Reports.
3. NET: Next European Torus (1985) Status Report, Commission of the European Communities, Directorate General XII, Fusion Programme, NET Report 51, Brussels.
4. ITER Conceptual Design Activities (1988/89) ITER documentation Series, published by the IAEA, Vienna.
5. Rubinacci, G. (1983) 'Numerical computations of the eddy currents on the vacuum vessel of a Tokamak', IEEE Trans. on Mag., Vol. Mag-19, N°6, pp.2478-2481.
6. Blum, J., Dupas, L., Leloup, C., Thooris, V. (1983) 'Eddy current calculations for the Tore-Supra Tokamak', IEEE Trans. on Mag., Vol. Mag-19, N°6, pp.2461-2464.
7. Albanese, R., Rubinacci, G. (1988) 'Integral formulation for 3-D eddy current computation using edge elements', IEE Proc. on Mag., Vol.135, Pr. A, N°7, pp.457-462.

8. Bossavit, A., Verite, J.C. (1983) 'The "TRIFOU" Code: solving the 3-D eddy currents problem by using H as state variable', IEEE Trans. on Mag., Vol. Mag-19, N°6, pp.2465-2470.

9. GRAZ COMPUMAG, Int. Conf. on Computation of Magnetic Fields (1987) Proc. published in IEEE Trans. on Mag., Vol.24.

10. Salpietro, E., Albanese, R., Coccorese, E., Martone, R., Mitchell, N., Rubinacci, G., NET: Next European Torus Operation Cycle, J. of Fusion Technology, Vol.14, N°1, 145.

11. Turner, L.R. (1988) Eddy Current Analysis in Fusion Devices, ANL Fusion Power Program Internal Report, ANL/FPP/TTI-223, Argonne (USA).

12. Crutzen, Y.R., Farfaletti-Casali, F. (1988) 'On the mechanical supports choice for NET first wall segments in the presence of transient electromagnetic interactions caused by plasma disruption events', Int. Symp. on Fusion Nuclear Technology (ISFNT), Tokyo (J).

13. CAPRI TEAM Workshop/Meeting (1988) Proc. published as an EUR Report N°12124 EN, CEC, JRC-Ispra.

14. BIEVRES TEAM Workshop/Meeting, Proc. published as an Eur Report N°12256 EN, CEC, JRC-Ispra.

15. Crutzen, Y., Rubinacci, G. (1989) 'Evaluation of the electromagnetic effects on a Tokamak first wall caused by a plasma disruption using a thin shell formulation', J. of Fusion Engineering and Design, Vol.11.

16. TOKYO COMPUMAG Conference on Computation of Magnetic Fields (1989) Proc. published in IEEE Trans. on Mag.

17. OXFORD TEAM Workshop/Seminar (1990) Proc. published as an EUR Report N°12988 EN, CEC, JRC-Ispra.

18. LONDON SOFT: Symp. on Fusion Technology (1990).

19. Albanese, R., Rubinacci, G. (1990) 'Formulation of the Eddy-Current Problem', IEE Proceedings, Vol.137, pt.A, N°1, pp.16-22.

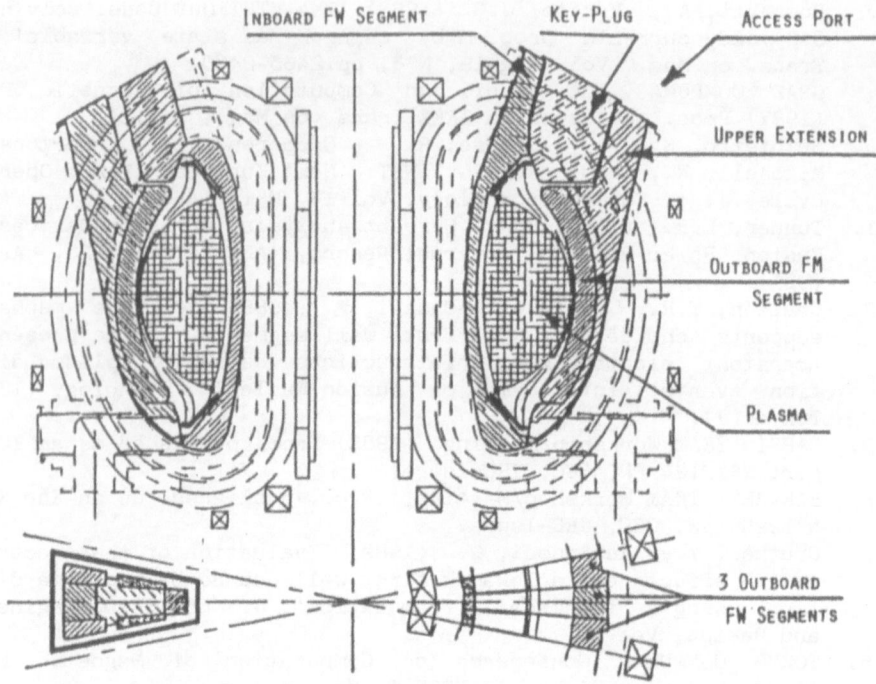

Fig. 1 - NET 1 reactor cross-section.

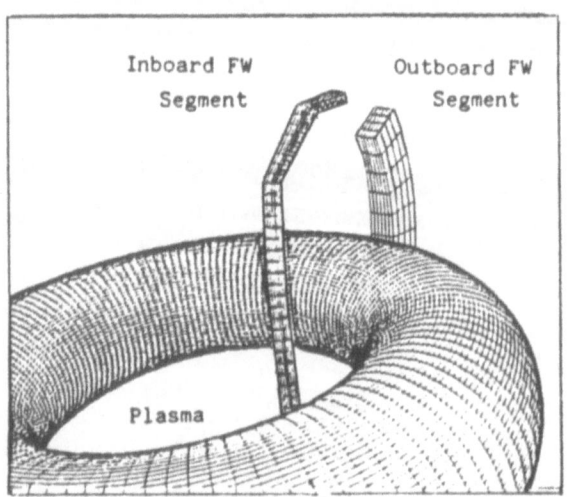

Fig. 2 - FEM model of an inboard and an outboard FW segment.

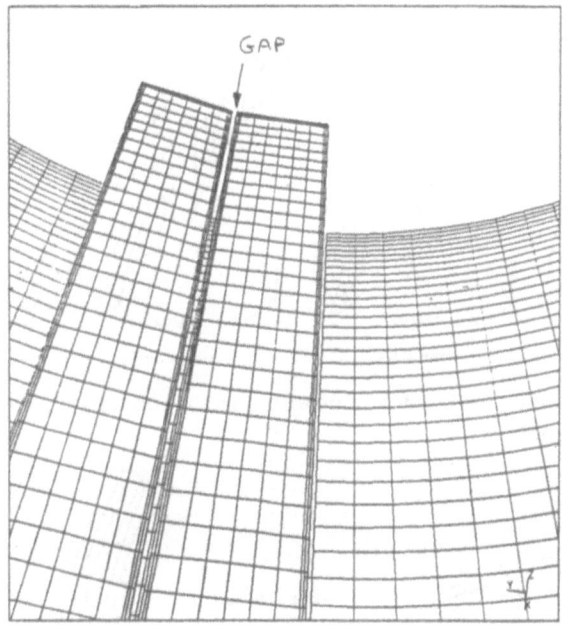

Fig. 3 – Real gap situation between adjacent segments.

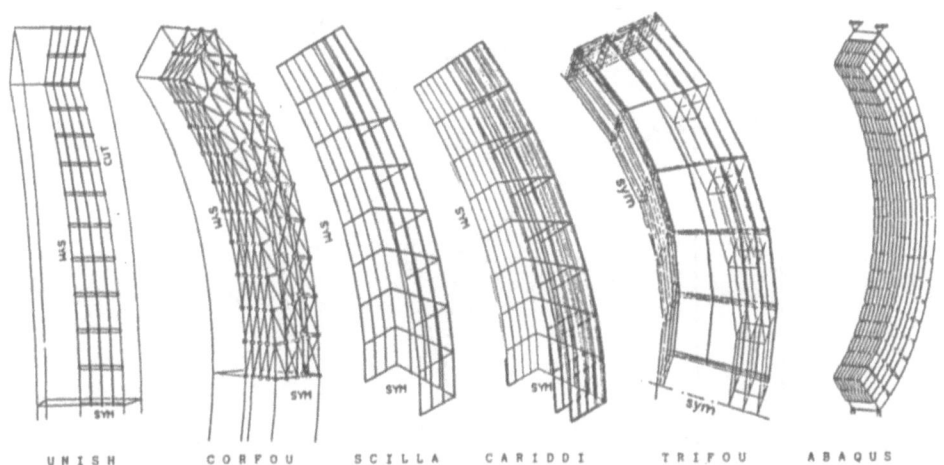

Fig. 4 – Finite element models of the outboard box-like first wall segment.

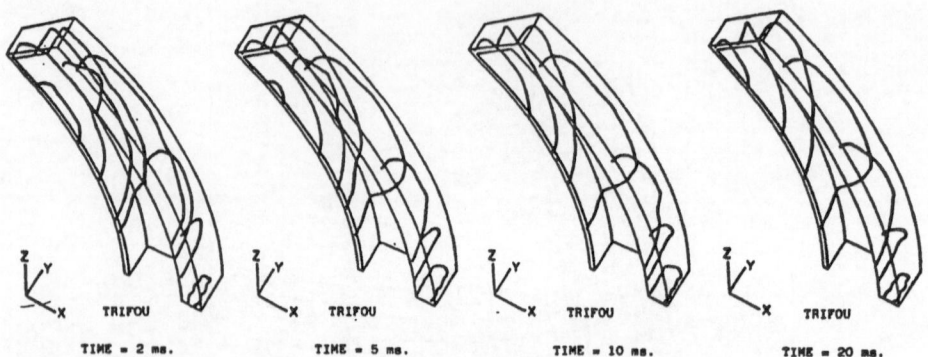

Fig. 5 – Time – dependent eddy current distribution (20 ms. plasma current quench).

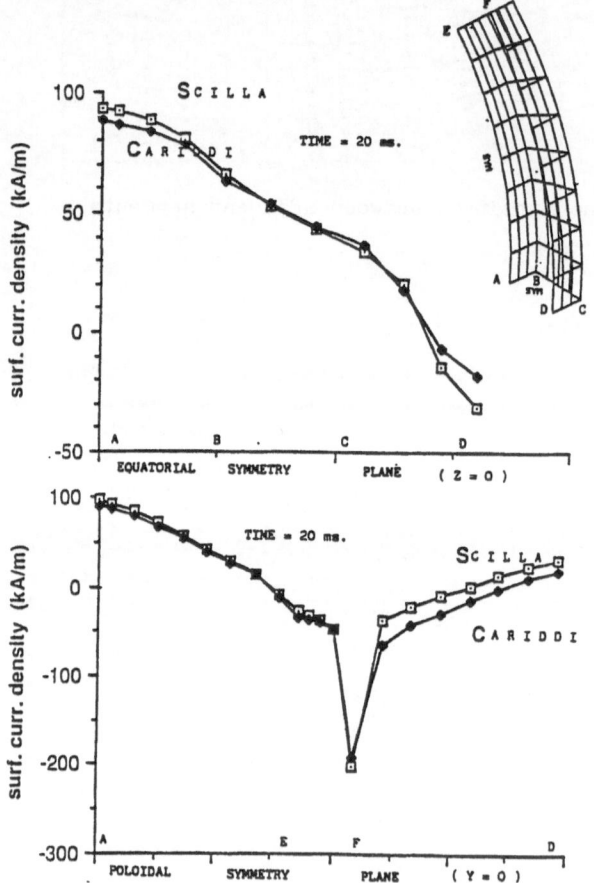

Fig. 6 – Eddy current variation in equatorial and polodial symmetry
planes.

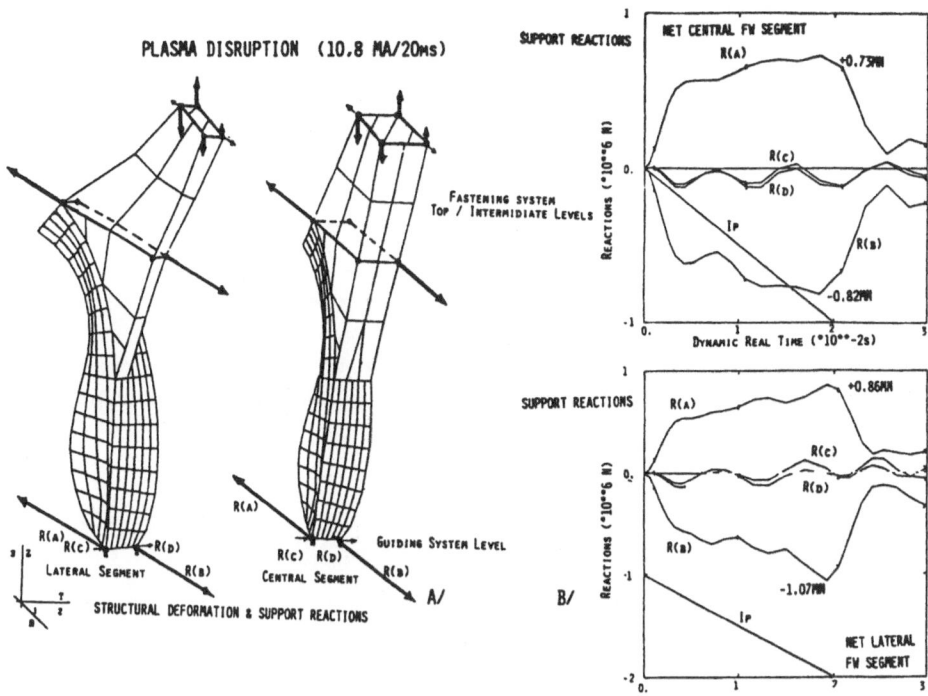

Fig. 7 - Outboard blanket segments under electromagnetic transients.

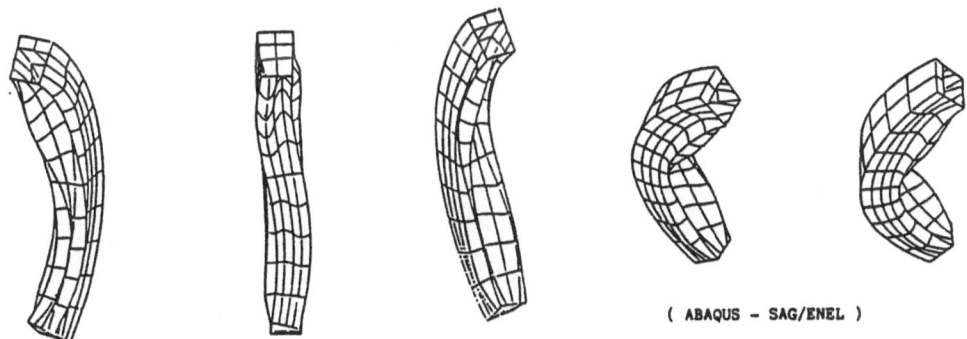

(ABAQUS - SAG/ENEL)

Fig. 8 - "Flashes" from the Animation Process During the Electromagnetic/Mechanical Transient.

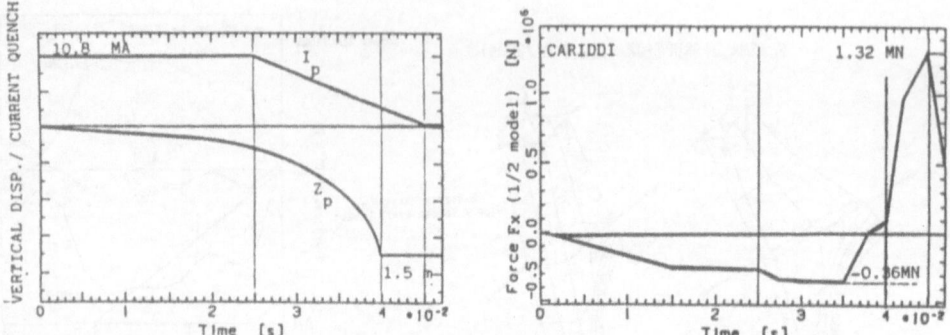

Fig. 9 – Plasma vertical motion and current quench histories radial
force resultant (Fx) during the dynamic plasma disruption.

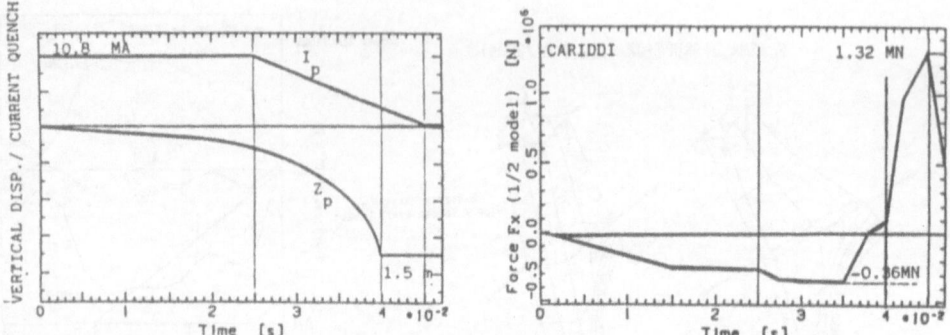

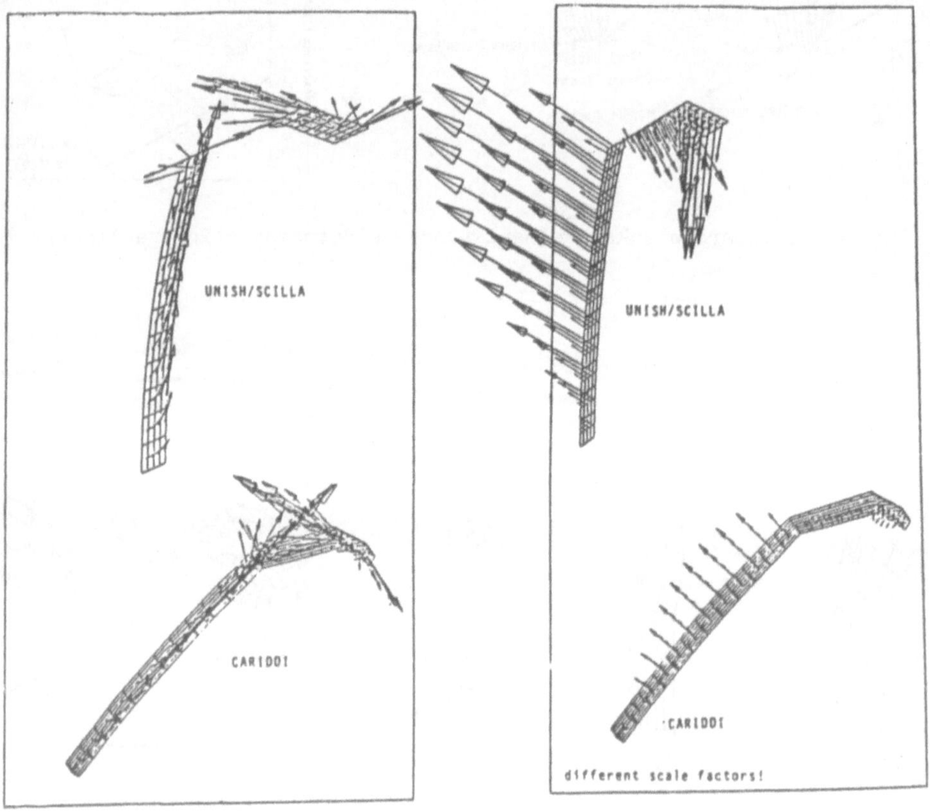

Fig. 10 – Eddy currents & electromagnetic forces distribution in
NET inboard first wall panel (1/4 model).

231

·	FW THIN SHELL THICKNESS. FOR FRONT SHELL, SIDE WALL	15 mm 30 mm
·	FW THICK PLATE THICKNESS. FOR RIGID BACK PLATE	50 mm
·	FW ELECTRICAL RESISTIVITY AT 316°C (AISI 316L):	0.9 μΩm
·	FW SEGMENT TOROIDAL ANGLE (TOT 32):	11.25°
·	TOROIDAL MAGNETIC FIELDS Bt AT PLASMA AXIS:	5.5 T
·	PLASMA CURRENT Ip:	18 MA
·	PLASMA CURRENT LINEAR DECAY TIME:	0.020 s
·	PLASMA MAJOR RADIUS:	5.5 m
·	NUMBER OF TETRAHEDRAL FINITE ELEMENTS (1/4 MODEL):	18120
·	NUMBER OF DEGREES OF FREEDOM IN THE MESH:	20705
·	TIME STEP INCREMENT (CRANK-NICHOLSON-INTEGRATION):	0.0002 s
·	EXTERNAL AXIALSYMMETRICAL CONDUCTORS (PF COILS):	6 × 2

Fig. 11 - ITER reactor - inboard/outboard segments -input parameters for inboard model (1/4 model using Trifou code).

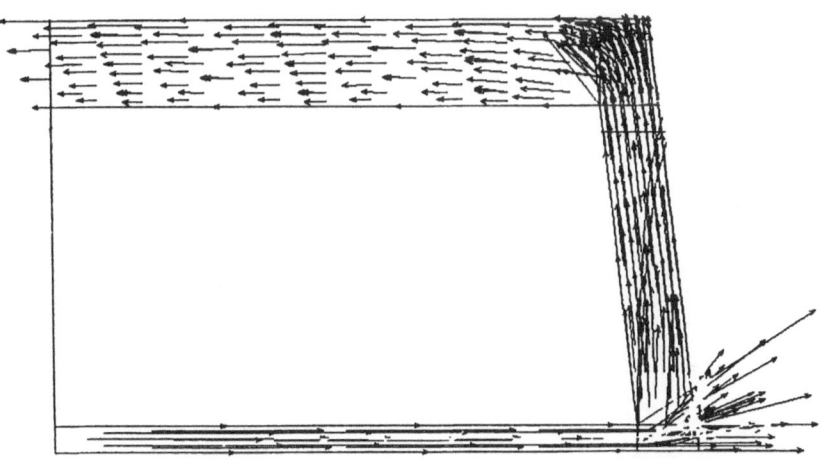

Fig. 12 - Eddy-Current Circulation in the equatorial region of the ITER inboard segment.

SOME RECENT DEVELOPMENTS IN 3-D TRANSIENT MAGNETOSTRUCTURAL COMPUTA-
TIONS IN THE FRAME OF THE   T E S L A   LABORATORY

S. Papadopoulos, N. Siakavellas, Y. Crutzen
Commission of the European Communities
Joint Research Centre
Institute for Systems Engineering and Informatics (ISEI)
21020 Ispra (Va) - Italy

1.   INTRODUCTION

The 3-D mechanical effects of transient electromagnetic-type phenomena
on conducting structures is still not completely investigated. In
fact, the numerical modelling of the mutual field-structure interaction must receive larger attention.

Recently, the Fusion Electromagnetic Induction Experiments
(FELIX) have been devoted to the study of transient magneto-structural
coupling in conducting plates, which correspond to the situation of
Tokamak in-vessel components where crossed steady and transient magnetic fields are superimposed. Up to now, the test problem 12, proposed by the TEAM (Testing Electromagnetic Analysis Methods)
Workshops, has been solved by simplified mechanical approaches; but,
as complex 3-D conductors are encountered in Tokamak devices, a fully
3-D coupled formulation is necessary for an accurate modelling.

Therefore, there is a great interest at JRC-Ispra to develop some
computational strategies in the field of real electromagnetic mechanical coupling and also to undertake experimental validation studies, in
order to confirm the effectiveness of the available numerical techniques.

2.   THE   T E S L A   ACTIVITIES

TESLA is the short term for Transient Electromagnetic/mechanical Simulation Laboratory. This laboratory has specific tasks related to /1/:
- Structural Integrity Assessment during Electromagnetic Transients;
- Validation of Computational Methodologies by Numerical Cross-Examinations and by Clean Experimental Measurements.

At present, detailed investigations concerning the effects of the
support system position on the global structure integrity of a Tokamak
first wall have been performed during the process of structural design
optimization /2/. Figure 1 illustrates the support position which min-
imizes the reference parameter of internal mechanical energy, for each
first wall segment, and the Von Mises stress intensity. In that spe-

Y. R. Crutzen et al. (eds.), Industrial Application of Electromagnetic Computer Codes, 233–238.
© 1990 *ECSC, EEC, EAEC, Brussels and Luxembourg.*

cific context, the following aspects need to be investigated in more detail:
- the body force distribution in multilayered shells including different material conductivities;
- the correct reproduction of the time-varying electromagnetic loading conditions during the stress analysis.

However, the real coupling between eddy currents and conductor motion has to be included in our electromagnetic-mechanical analyses, since there is no quantitative answer about how much the coupling would decrease the expected deflections and stresses.

Therefore, a dual approach for TESLA has been defined in order to extend our knowledge in that context. First, various numerical modelling/analysis efforts in eddy current analysis have been carried out. Then, an experimental test facility has been designed for the first validation studies.

2.1 Numerical Activities

During the past few years, thermomechanical (MECH) software tools have been developed and applied for nonlinear transient dynamic analysis /3/. Recently, the activity has been devoted to the application of numerical methods implemented into electromagnetic (EM) computer tools for eddy current analysis /4/.

In this way, validation studies are initiated performing the same simulations with different codes in order to compare their capabilities /5/. The validation activity is presently focused on the NET (Next European Torus) and ITER (International Thermonuclear Experimental Reactor) design, in order to assess the behaviour of the in-vessel plasma-facing components subjected to transient electromagnetic effects and to solve the related structural issues.

For this purpose it is planned to complete the present numerical simulation activity with the creation of an experimental test facility.

2.2 The Experimental Test Facility

The purpose of the experimental test facility is the verification and validation of the computational methodologies during magnetostructural transient events. The technical issue to be addressed is related to some in-vessel components, particularly the divertor plate modules, subjected to reference plasma disruption scenarios (rapid current quench in the plasma ring, creating a strong vertical magnetic field variation).

In a first phase a small prototype machine is considered, the test-rig ELBA (ELectromagnetic BAsic experiment). The experimental test-rig creates transient electromagnetic conditions by means of crossed magnetic fields (steady and fast-time varying). Performing small-scale clean electromagnetic experiments, this test-rig (Figure 2) will permit to validate, up to some extent, the computer tool applications and to define some benchmark problems (to be proposed to the TEAM Workshop Planning Board). Some design parameters are: minimum inner solenoidal diameter: 250 mm, available test region volume: 100x100x200 mm^3, vertical dipole field decay: 50 T/s, ratio between the solenoidal straight field and dipole peak field intensities: about 5, allowed flux disuniformity: a few %.

3. NUMERICAL SIMULATIONS RELATED TO ELBA EXPERIMENTS

The numerical simulation activities are actually oriented into the following two fields:
- Electromagnetic (EM) computations of simple small-size models;
- Simulations of electromagnetic/mechanical (EM/MECH) coupling.

3.1 Electromagnetic Computations of Simple Small-size Models
Numerical calculations of simple small-size models simulating flexible conducting components have been made in order to determine the experimental parameters of the ELBA experiment /6/. The models considered for these calculations are a rigid square frame and three thin plates. The plates provide a simple model for the limiter blade of a Tokamak fusion reactor. The choice of the dimensions of the models has been made respecting the limitations of the available test region volume. Various simulations of vertical transient magnetic fields Bz(t) were also applied in order to investigate an appropriate field decay time. The first results show that the vibrations and deflections of the small-size plate models, in both bending and torsional modes, can be measured by means of standard LASER techniques.

3.2 Simulations of Electromagnetic/Mechanical (EM/MECH) Coupling
In a Tokamak, interactions between the eddy currents and the magnetic fields (toroidal and poloidal) cause mechanical effects (deflection and stress) in the passive conducting components. The structural component, as it deforms, intercepts additional magnetic flux and complementary eddy currents, circulating in opposite direction, are generated. This coupling effect between deflection and eddy currents can mitigate the potential damages to the structures predicted by performing independent (uncoupled) electrical and structural analyses.

In that context, the interface software program TRINACRIA /7/ has been developed. TRINACRIA provides a semi-automatic iterative process for the evaluation of corrective terms (modifying the eddy currents). Actually, it acts as an "external" interface between the electromagnetic program CARIDDI /8/ and the mechanical program ABAQUS /9/.

On the other hand, in order to obtain a first estimate of the deflection of the small plate models, for the ELBA experiments and considering the EM/MECH coupling effects, the program TRIFOU /10/ has been applied with a small hole in the middle of the plate. Starting from the steady-state version, the real and imaginary part of the total current circulating the plate has been deduced. Then, the coupled EM/MECH equations of the plate have been solved numerically and the time variations of the total current and deflection of the plate have been determined, for two different materials (aluminium and copper). In the case of the copper plate, two different horizontal magnetic fields have been considered: Bx = 0.1 T and Bx = 0.5 T. As shown in

Figure 3, the magnetic rigidity effects are well pronounced with a constant field of O.5 T.

4. CONCLUSIONS

A numerical assessment of simple small-size models has been carried out, in order to choose the best experimental parameters for the ELBA test-rig and the specimens. In relation with the different magnetic and mechanical time constants, investigations must be extended in order to obtain an optimal EM/MECH coupling and a pronounced magnetic rigidity effect.

In the presence of abnormal electromagnetic transients in fusion devices, additional investigations can be performed in order to evaluate the effective drop of the induced electromagnetic loads obtained by an increased flexibility to motion (i.e. the first wall panels) and into the need of support systems including fastening and attaching locks (i.e. locking belt systems).

5. REFERENCES

1. Crutzen, Y., Biggio, M., Castillo, E., Farfaletti-Casali, F., Nervi, M., Papadopoulos, S., Siakavellas, N. (1990) 'TESLA, A Transient Electromagnetic/Mechanical Simulation Laboratory', Oxford Team Workshop Proceedings, CEC, EUR Report N°12988 EN.
2. Crutzen, Y., Farfaletti-Casali, F. (1989) 'Structural design problems related to plasma disruption events of the removfable NET first wall segments', Tokyo-ISFNT Symposium, publ. Nuclear Engineering and Design/Fusion Journal, Vol.9, pp.101-105.
3. Crutzen, Y., Russo, G., Youtsos, A. (1989) 'Development of a UMAT subroutine for viscoplastic material behaviour and 3-D numerical analysis of the Tokamak structural dynamic response', ABAQUS User's Conference Proceedings, Stresa (I).
4. Crutzen, Y. (1990) 'Electromagnetic analysis applied to the Tokamak first wall design', Companion paper.
5. Crutzen, Y.R., Farfaletti-Casali, F. (1990) 'Electromagnetic validation studies needed for the design of net fusion devices', COMPEL, The Int. Journal for Computation and Mathematics in Electrical and Electronic Engineering, Vol.9, Suppl.A, 97-100.
6. Papadopoulos, S. (1990) 'Optimization of clean experimental models for the TESLA small-size test-rig', Int. Technical Note, CEC, JRC-Ispra, N°I-90-xx.
7. Siakavellas, N. (1990) 'TRINACRIA, Interface Program in Magneto-Structual Analysis', Int. Technical Note, CEC, JRC-Ispra, N°I-90-yy.
8. Albanese, R., Rubinacci, G. (1990) 'Formulation of the eddy current problem', IEE proceedings, Vol.137, pt.A, N°1, pp.16-22.
9. Hibbitt, H.D., Karlsson, B., P. Sorensen Inc. (1989) 'ABAQUS Manuals: Vol.1: User's Manual, Vol.2: Theory Manual, Vol.3: Example

Problems Manual, Vol.4: Systems Manual', Version 4.8, Providence (RI), USA.

10. Bossavit, A., Bidan, J.Y., Cahouet, J., Chaussecourte, P., Lamaudière, J.F., Rose, C., Vérité, J.C. (1990) 'Approche théorique et pratique à l'utilisation du logiciel TRIFOU', Cours TRIFOU, EdF (Electricité de France)/DER/IMA/MMN, Clamart (F).

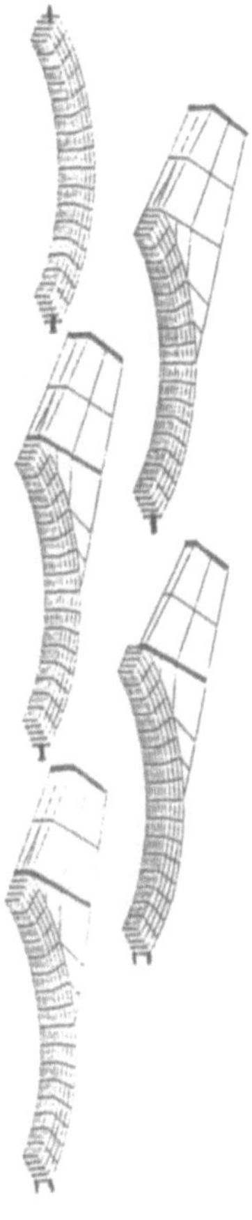

TYPE	MAX REACTION FORCES			MAX STRESS	STRAIN ENERGY
	Level	Direct.	Value [t]	Von Mises [MPa]	[kJ]
	MED	Z	19		
C	MED	T	40	190	332
	PIN	T	40		
	SUP	Z	74		
C	SUP	R	66	130	261
	PIN	T	41		
	SUP	Z	30		
C	MED	R	51	125	211
	PIN	T	42		
	SUP	Z	21		
L	MED	R	85	120	151
	PAT	T	107		
	SUP	Z	15		
C	MED	R	44	110	101
	PAT	R	82		

LEVEL OF FASTENING AND GUIDING SYSTEMS:
SUP : Upper Fastening
MED : Intermediate Fastening
PAT : Bottom Twin Pin Guiding
PIN : Bottom Single Pin Guiding
C : Central Outboard Blanket ; L : Lateral Outboard Blanket
PLASMA DISRUPTION (10.8 MA / 20 ms)

Fig.1: Effects of the support position on the structural integrity of the NET first wall outboard segments during a major plasma disruption (10.8 MA/20 ms plasma quench).

238

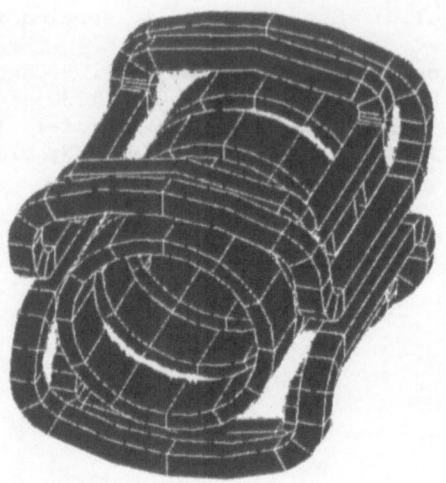

Fig.2: Dipole and solenoid magnets configuration for the ELBA test-rig.

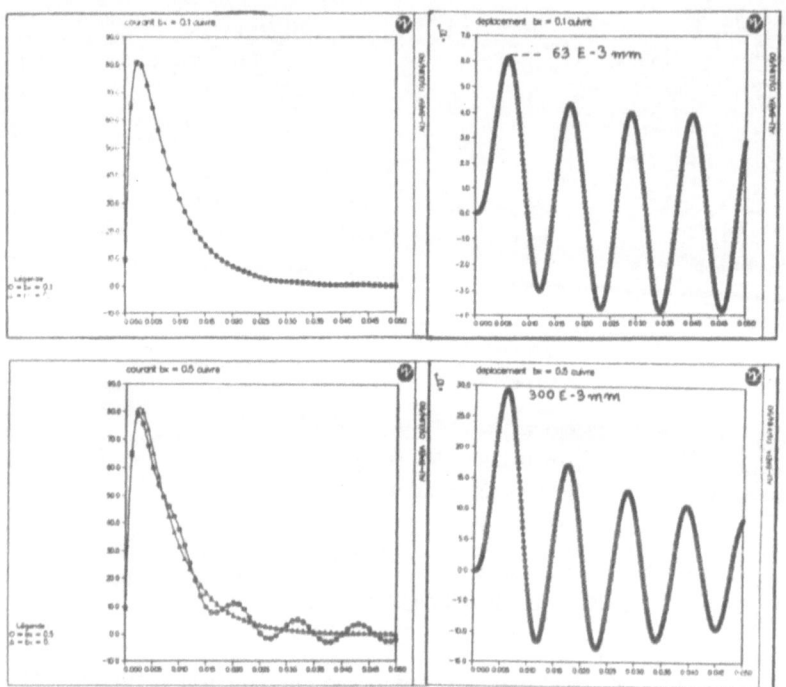

Fig.3: Current and deflection variations as a function of time (a: Bx = 0.1 T, b: Bx = 0.5 T, copper).

NUMERICAL APPROACH IN MODELING THE ELECTRIC ARC

J. Schneider
CEC Joint Research Centre, 21020 Ispra, Italy

ABSTRACT. This paper describes a PC-based calculation tool for the
solution of systems of ordinary differential equations, suited for a
wide range of problems encountered in scientific applications. The approach is demonstrated for the analysis of high power electrical arcing in a conceptual compact high-field tokamak.

INTRODUCTION

Examination of stored energies and potential pathways of energy release indicates that magnet accidents could be the source of damage to
the magnets themselves and to other subsystems of the device. For a
high-field tokamak (HFT) a magnetic flux density of about 10 T is
foreseen in the toroidal field. Integrated over the toroidal volume,
the energy stored in the toroidal magnetic field is in the order of
several hundred MJ. If only part of this energy is released during an
accident a release of radioactive materials to the containment and/or
the environment may result. It is known from experience with large
magnets in high-energy-physics laboratories, that most of the magnet
accidents have finally resulted in arcing. Arcing therefore
constitutes a main safety concern.
　　　　Designers are therefore concerned about potential consequences of
arcing accidents to find a basis for guidelines to be followed in the
construction-, licensing-, operation- and maintenance-phases of the
experiment.
　　　　Extrapolation of the existing experience to HFTs seems to be
inadequate, because here the magnetic field energies are 2-3 orders of
magnitude higher. Theoretical analysis is therefore performed, based
on the model developed at Ispra /1/. Input data requirements of the
model are primarily current-voltage characteristics of the arc in realistic fusion environments (high currents and high magnetic fields).
Subprograms have been developed regarding inductance and magnetic
field energy of the coils. Further development is envisaged regarding
the coupling of the model with heat transfer and structural analysis
programs.

Y. R. Crutzen et al. (eds.), Industrial Application of Electromagnetic Computer Codes, 239–246.
© 1990 ECSC, EEC, EAEC, Brussels and Luxembourg.

SAFETY ISSUES

The easiest fusion reaction to achieve is between D and T the latter being radioactive (half life 12.3 years). Furthermore the released neutrons are absorbed in the reactor structure and create substantial inventories of radioactive activation products. The bulk of the radioactive products is trapped in the solid structure material and can not be dispersed as such. However small amounts of radioactive dust inside the torus (produced by erosion due to plasma wall interactions) need special attention. Especially during maintenance and accidental conditions. An additional concern for fusion plant safety are the energies stored in the device. Of special concern for HFTs is the magnetic field energy stored in the magnets, especially in the toroidal field coils.

MAGNET ACCIDENTS

If only part of the magnetic field energy is released accidentally, damage to the magnets and to other subsystems of the device may result. Magnet accidents may be caused by imbalanced forces, shorts and arcs. It is known from magnet operating experience, accumulated in high-energy physics laboratories, that most of the magnet accidents have finally resulted in arcing. The arcing process constitutes therefore a main safety concern. An extrapolation of the existing experience seems inadequate because the stored magnetic field energy in HFTs is two orders of magnitude higher. Theoretical analysis has therefore been initiated.

THEORY OF THE ARC DISCHARGE

The electric arc is one of the phenomena of great importance for the practical operation of large scale fusion magnets. No closed mathematical form has been presented in any of the arc theories so far published. The microscopic processes involved elude mathematical analysis mainly because of lack of data.

A theoretical model based on an energy balance of the basic physical phenomena has therefore been set up. This allows for first basic estimates of arcing consequences. Because of the uncertainty in the current-voltage characteristic of the arc in realistic fusion magnet environments, the results of the model must be considered as order of magnitudes.

The conversion of stored magnetic field energy to heat and kinetic energy is given by the following power balance:

$$LI \frac{dI}{dt} = -RI^2 - nVarcI - IBlv \sin \phi \qquad (1)$$

```
I    = current (A)
L    = inductance (Hy)
R    = resistance (Ω )
Varc = arc voltage (V)
Φ    = angle between direction of current and magnetic field
n    = number of arcs generated
l    = length of missile (m)
v    = velocity of missile (m/s)
B    = magnetic flux density (T)
t    = time (s)
```

Classically the voltage of electrical arcs is given by AYRTON's formula:

$$Varc = a + b + \frac{c+ds}{I} \tag{2}$$

s is the arc length. The values a, b, c, d are taken from measurements reported in the literature /2/.

Because of the self pinching effect in high current arcs this formula might not be valid. New experimental information is needed for high power metal vapor arcs, which must be expected in fusion magnet environments.

The mechanical power developed by a missile through acceleration by the LORENTZ force is given by:

$$mv \frac{dv}{dt} = IBlv \sin \Phi \tag{3}$$

m is the missile mass. From (3) the equation of motion of the missile, moving under the LORENTZ force alone, is obtained:

$$\frac{d^2s}{dt^2} = \frac{IB \sin\Phi}{\rho A} \tag{4}$$

with:
A = cross section of missile (m^2)
ρ = density of missile material (kg/m^3)
s = distance travelled by missile (m)
v = velocity of missile (m/s)

INTEGRATION OF SYSTEMS OF ORDINARY DIFFERENTIAL EQUATIONS

All numerical integration techniques are based either on Taylor- series expansion or on polynomial approximations. Both methods give exact solutions if the dependant variable is a polynomial of finite degree. Among standard numerical methods are forward Euler, backward

Euler, trapezoidal rule, and Runge-Kutta methods. Special techniques must be applied for stiff systems. Suitable methods for stiff systems are backward Euler, trapezoidal rule, and implicit Runge-Kutta methods.

The problem is formulated in vector form /4 /:

$$y'(x) = F(x,y) \tag{5}$$

with:
y = (y1,y2,.......yn)
F = (f1,f2,.......fn)

The initial value problem is defined by:

$$y(x=0) = A \tag{6}$$

with:
A = (a1,a2,........an)

A 4th order Runge-Kutta algorithm has been chosen because of accuracy, computational stability and speed of integration. It requires only four function evaluations for each integration step, and does not require prior evaluation of higher derivatives as the Taylor-expansion of order 4 would imply.

The algorithm is reproduced in the following:

$$k1 = hf(x,y) \qquad)$$

$$k2 = hf(x+h/2,y+k1/2) \qquad)$$

$$k3 = hf(x+h/2,y+k2/2) \qquad) \tag{7}$$

$$k4 = hf(x+h,y+k3) \qquad)$$

$$y(i+1) = y(i) + \frac{1}{6} (k1 + 2k2 + 2k3 + k4) \}$$

Note that y(0) is known and that the solution of the system of differential equations proceeds in steps of h from y(i) to y(i+1).

PC-BASED COMPUTER PROGRAM FOR THE FOURTH ORDER RUNGE-KUTTA ALGORITHM

A computer program has been developed in BASIC which integrates a system of n first order ordinary differential equations, using the RUN GE-KUTTA method. This approach is highly interactive and user friendly and is not limited by mainframe running costs. The main program contains 3 subroutines for initialization, calculation of the derivatives, and the fixed step standard RUNGE-KUTTA integration routine. A special subroutine enables the calculation of the results when analytic solutions are possible.

The system of differential equations (5) must be specified by the user. Before entry the y(i) must contain the initial values of the problem. This step is performed interactively. On exit the y(i) contain the computed values of the solution at the final value of x.

Besides the specific differential equations used and the integration method, accuracy depends on the step size. It is controlled by a step size control parameter. For stiff problems with rapidly varying components the user may reduce the step size.

TRANSIENT ANALYSIS OF ELECTRICAL ARCING IN A HFT PR-MAGNET CIRCUIT

A HFT is a toroidal closed configuration in which the plasma particles are confined by nested surfaces composed of helical magnetic field lines. The magnetic field is principally composed of three parts. The strongest component is the toroidal field TF, created by a toroidal solenoid magnet. The poloidal field is created by the toroidal plasma current and additional poloidal field coils PF, surrounding the plasma. The toroidal plasma current is induced by creating a magnetic flux change through the center of the device, which creates a toroidal electric field. The plasma current serves the function to create a component of the poloidal field and additionally heats the plasma. The coils for creating the magnetic flux change are called Ohmic heating coils OH.

HFTs have been proposed for several years as a means to achieve ignition in a small compact device at acceptable costs. Their principle features are /3/:
- The centering forces on the toroidal field coils are reacted internally, out of plane forces are reacted externally.
- Axial precompression of the toroidal field coils by means of an electromagnetic press PR.
- The poloidal field coils are external to the toroidal coils.
- Magnetic fields are produced by Cu coils operating at LN2 temperature.

An electrical arc discharge in the PR-magnet circuit of a HFT may be the result of a voltage breakdown, conductor fracture or conductor melting. These events are associated with the generation of high power arcs, driven by the energy stored in the magnetic field.

Main parameters of a HFT PR-coil are outlined in table 1.

Table 1. Main HFT PR coil data.

Current	112500	A
Current density	1.25e8	A/m2
System inductance	7.58	mH
Resistance	0.0035	V/A
Stored Energy	48	MJ

For an electric arc in series with the magnet system the following differential equation applies for the discharge:

$$L \frac{dI(t)}{dt} + RI(t) + Varc = 0 \tag{8}$$

I = current (A)
L = inductance (Hy)
R = resistance (V/A)
$Varc$ = arc voltage (V)
t = time (s)

For constant arc voltage the above inhomogeneous first order ordinary differential equation may be solved analytically.
The solution of the homogeneous equation is:

$$Ih(t)' = I(t{=}0)e^{-\frac{Rt}{L}} \tag{9}$$

The solution of the inhomogeneous equation (8) can be obtained in terms of an additional integration by setting:

$$I(t) = constIh(t) \tag{10}$$

Substitution into (8) and integration yields the general solution of (8):

$$I(t) = \left(const - \frac{Varc}{RI(t{=}0)} e^{\frac{Rt}{L}}\right)I(t{=}0)e^{-\frac{Rt}{L}} \tag{11}$$

With the initial value $I(t{=}0)$ substituted into (11), the solution of (8) is:

$$I(t) = I(t{=}0)e^{-\frac{Rt}{L}} - \frac{Varc}{R}\left(1 - e^{-\frac{Rt}{L}}\right) \tag{12}$$

By setting $I(t){=}0$, a theoretical burn time of the arc discharge is obtained:

$$bt = -\frac{L}{R}\ln \frac{1}{\frac{I(t{=}0)R}{Varc} + 1} \tag{13}$$

RESULTS OF A TRANSIENT ANALYSIS

Measurements from the TESPE experiment /5/ have shown that the classical AYRTON formula for the current-voltage characteristic of the electrical arc is not valid for arc currents higher than 1000 A. This is due to the self-pinching effect of the arc column and the high pres sure in the discharge region. In the absence of detailed measurements for the specific HFT design, the arc voltage is varied between 30 and 200 V. For the HFT PR magnet circuit the main discharge results are outlined in table 2.

Table 2. HFT PR magnet discharge results.

Varc(V)	bt(s)	E(MJ)	P(MW)	En(%)
30	5.73	5.83	1.02	12
50	4.73	8.80	1.86	18
70	4.10	11.32	2.76	24
100	3.46	14.48	4.18	30
120	3.15	16.28	5.17	34
150	2.79	18.62	6.67	39
200	2.36	21.80	9.25	45

with:
bt = arc burn time (s)
E = energy dissipated in the arc at bt (MJ)
P = mean arc power (MW)
En = percentage of energy dissipated in the arc (%)

It can be seen from table 2 that the arc burn time is in the range of seconds, whereas the normal discharge time constant would be of the order of some tenth of seconds. Therefore the presence of an arc in the discharge circuit tends to increase the discharge time, leading to a high destruction potential. It can also be seen that this effect decreases with increasing arc voltage. A better knowledge of the arc voltage for the specific HFT design would therefore be of great value.

Since the arc develops a power of several MW the mechanical yielding of the magnet structure and the formation of missiles may occur in the outlined time scales.

MISSILE VELOCITIES

Missile velocities were calculated with the hypothetical assumption that one conductor of length 1 m breaks out of the PR magnet. It is accelerated in a constant magnetic field of magnetic flux density 7 T and with arc voltages varying in the range of 30-200 V. If the data are arbitrarily interrupted at about 0.5 m distance travelled, the missile would reach a velocity in the order of 100 m/s in some ms. No influence of the arc voltage was observed, indicating that the missile moves relatively independent of the current-voltage characteristic of the arc during the first short time intervals. Since active decoupling of the energy from the arc in this short time scales is difficult to achieve, multiple current arcs may emerge. The development of multiple current arcing depends on the movement of the high power arc in the magnetic field. The single arc column may not move in the classic direction as expected from the current and magnetic field directions, but may show retrograde motion as already observed for low current arcs /6/. Further experimental work should be directed towards this phenomenon. Under the outlined assumptions arcing results constitute estimates to those expected in real situations. In further steps the assumptions may be successively eliminated and confronted with experimental results.

REFERENCES

1. H. Schneider, A. Caretta, "Missile Generation due to Electrical Arcing in High Field Fusion Magnets", Proceedings, 9th Symposium on Engineering Problems in Fusion Research, Chicago, October 1981.
2. W. Finkelnburg, H. Maecker, "Elektrische Bögen und Thermisches Plasma", Handbuch der Physik Bd.22, Springer Verlag Berlin, 1956.
3. B. Coppi, "Physics and Technology of Ignition Experiments", Proceedings, Symposium on Fusion Engineering, Austin, November 1985.
4. R. Zurmühl, "Praktische Mathematik für Ingenieure und Physiker", 9. Auflage, Springer Verlag Berlin, 1965.
5. Ergebnisbericht über Forschungs- und Entwicklungsarbeiten, Institut für Technische Physik, KFK 4545, 1988.
6. A. Eidinger, W. Rieder, "Das Verhalten des Lichtbogens im transversalen Magnetfeld", Archiv für Elektrotechnik, 43, 1957.

APPLICATIONS OF ELECTROMAGNETIC ANALYSIS TO NON DESTRUCTIVE TESTING

K. Betzold
Institut für zerstörungsfreie Prüfverfahren
Fraunhofer Gesellschaft
Universität Gebäude 37
6600 Saarbrücken 11
BRD

ABSTRACT. This paper outlines the potential and the qualification of several categories of electromagnetic models for the nondestructive inspection with electromagnetic methods; these are analytic 2-d FE and 3-d FE models.
The selected applications refer to
- eddy current inspection of nuclear components like steam generator tubing and heavy plates
- microstructure characterization of steels by σ, μ-measurement (σ = electrical conductivity, μ = magnetic permeability)
- magnetic particle inspection of automobile components

1. INTRODUCTION

According to J.M. Coffey /5/ there are 5 major areas in eddy current ndt where models are making a contribution:

1.1 OPTIMIZED CAD OF INSPECTION SYSTEMS

- coil design (geometry, windings)
- coil type (absolute-, differential-, transmitter-receiver-, pick-up-, bobbin coil)
- test frequencies

1.2 VALIDATION OF PROCEDURES

- simulating the effectiveness of an inspection procedure at detecting and evaluating the defects of concern.

1.3 INTERPRETATION OF INSPECTION SIGNALS

- disturbing signals
- defect signals
- iterative forward computation (signal prediction)
- link to expert system

Y. R. Crutzen et al. (eds.), Industrial Application of Electromagnetic Computer Codes, 247–263.

1.4 EXTENDING THE RANGE OF APPLICATION OF NDT TECHNIQUES

- horizontal diversification (f.e. medical technique, not discussed here)

1.5 EDUCATION AND TRAINING

- deeper understanding of eddy current phenomena improves the inspection technique and signal interpretation

The applications within this paper will support 4 of these items (except 'Extending the Range').

2. DISCUSSION OF ELECTROMAGNETIC MODELS

Which tools are available to model electromagnetic ndt-phenomena? Characteristic configurations for 3 categories of modelling tools are shown in fig. 1.

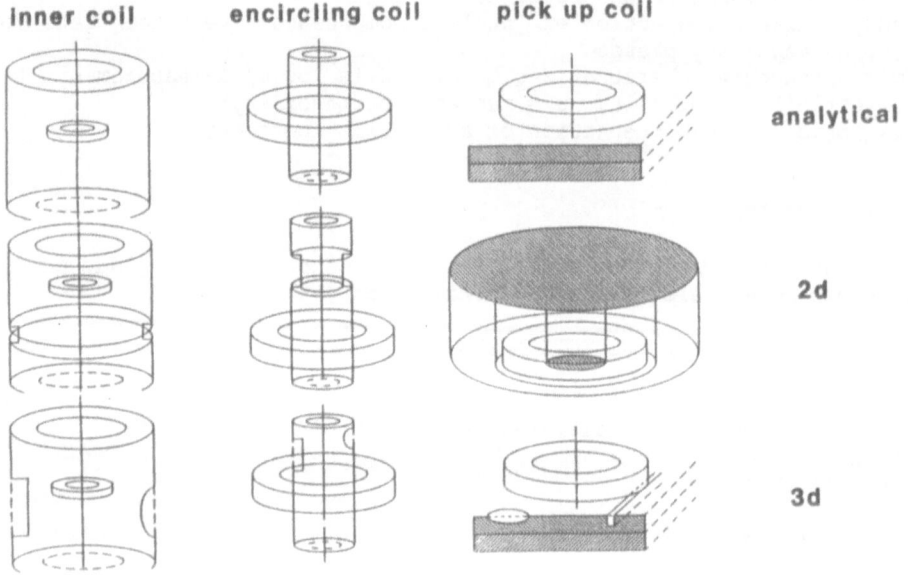

Fig. 1: Modelling of eddy current inspection problems

Row 1 refers to integral expressions, developed by C.V. Dodd /1/. They can be used to simulate thickness-, distance-, conductivity- and permeability measurement at multi-layered test objects. In tube inspection with inner or encircling coils only material interfaces in axial direction can be handled (reason: separation of variables). The Dodd-model includes real inspection coils (absolute-, differential-, transmitter-receiver-, pick-up coil).

The numerical evaluation of the integral expressions takes less than a minute per impedance point on a PC; Dodd-results represent a reference for codes using discretizations.

Row 2 of fig. 1 corresponds to typical axisymmetric inspection problems that can be tackled with finite elements. These allow to simulate the dynamics of the inspection by predicting signal loci curves.

Defects like slots with a certain width and depth or disturbing influences like tube sheet or roll expansion or sludge deposit - all assumed to be axisymmetric - can be handled separately or superimposed with each other.

The discretization into finite elements has to take care of the electrical and magnetic material properties and the inspection frequency; discretization errors have to be detected. The computing time for one point in the impedance plane is several minutes on a PC. Row 3 of fig. 1 reflects some 3-d configurations of eddy current ndt; signal loci curves of real defects can only be predicted with 3-d modelling facilities. In 3-d the use of a preprocessor to generate the test configuration and of a postprocessor to display the results is indispensable. Nevertheless the chance of discretization errors or misinterpretations of the signal is bigger than in 2-d.

Because of the memory and CPU they need, most of the 3-d codes have been developed for mainframe machines. Due to the increasing capability of the workstations in the last years some codes can be run on workstations.

The computing time of a 4000 element discretization (ref. fig. 12) is typically 2 hours CPU on a VAX 3100 for one impedance point.

3. APPLICATIONS

3.1 PENETRATION DEPTH OF EDDY CURRENT COILS

If a subsurface defect, starting in the interface between cladding and base material of a pressure vessel has to be detected it is necessary to provide a current density in that depth which is sufficient to result in a variation of the coil impedance. The major influences determining the current density are the electrical and magnetic properties of the test object, the inspection frequency and the coil dimension. For a rough overview upon the penetration depth it is quite common to use the well-known formula for the "standard penetration depth"; but has it been proved that this formula is valid for real coils?

This question has been investigated by evaluating Dodd's integral expressions.

material	coil dimens. / frequency	O ⌀10mm coil3	O ⌀20mm coil2	O ⌀30mm coil1	standard pene- tration depth
austenite	.16kHz	8.25%	12%	22%	40mm
$\mu = 1$ $\sigma = 1MS/m$	9.5kHz	52%	68%	81%	5.16mm
ferrite	.16kHz	80%	92%	98%	2.5mm
$\mu = 50$ $\sigma = 5MS/m$	9.5kHz	94%	97%	~100%	0.3mm

Fig. 2: Penetration depth of pick-up coils

In fig. 2 the penetration depth (1/e-decrease of the current density at the surface) of 3 pick-up coils of different size is compared with the standard penetration depth for 2 single-layered materials - austenite and ferrite - and for 2 frequencies. The largest difference occurs to the austenitic material; if it is inspected with .16 kHz and a coil having a diameter of 10 mm the penetration depth is only 8.25% of the standard penetration depth. for 9.5 kHz it is still not more than 52%.

The differences for the ferromagnetic test objects are much smaller. The same coil as above (10 mm ⌀) provides a penetration depth of 80% at .16 kHz and 94% at 9.5 kHz. For a large coil like 30 mm in diameter, the differences are negligible.

Argumentations based on standard penetration depth therefore may be able to give a quick idea of the real coil penetration depth but for the investigations of an electrical engineer they are not acceptable.

The eddy current penetration into a steam generator tube and into a full rod of the same material excited by an encircling coil is discussed in fig. 3. The important result is, that the current density - compared in the same depth - is considerably higher in the tube than in the rod. At 200 kHz the current density in a depth of 1.2 mm is 66% of the surface value for the tube and 35% for the rod. At 400 kHz it is

43% for the tube and 22% for the rod and at 800 kHz still 21% for the tube and 11% for the rod.

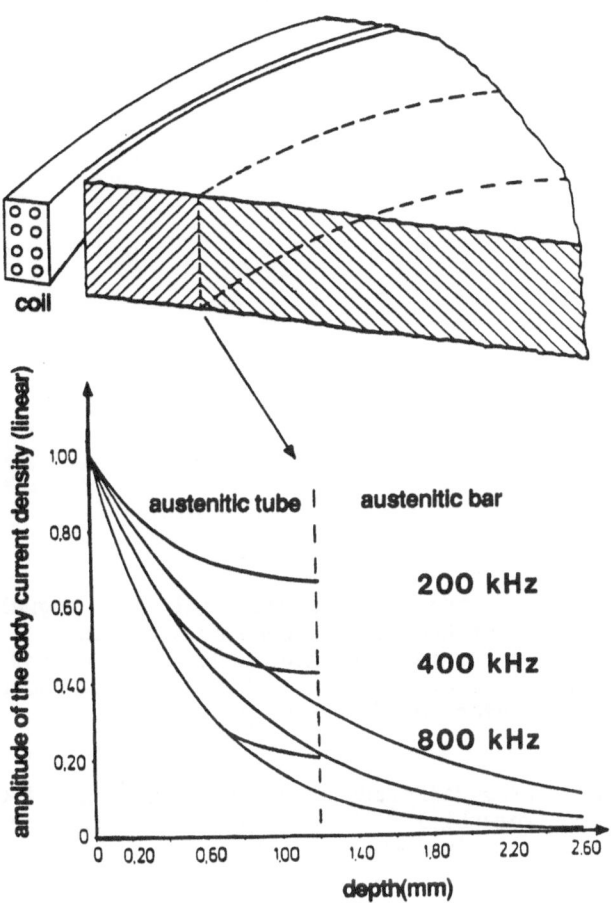

Fig. 3: Eddy current penetration in tubing and bars

3.2 COIL DESIGN IN STEAM-GENERATOR TUBE INSPECTION

The goal of these investigations was to optimize the design of differential bobbin coils for detecting and assessing circumferential slots on the outside of the tube. In fig. 4 the influence of the distance between the windings is demonstrated for a circumferential slot having 0.5 mm in width 40% wall reduction. The tool used for these investigations is a 2-d FE-code developed by R. Palanisamy /2,3/ under EPRI-contract.

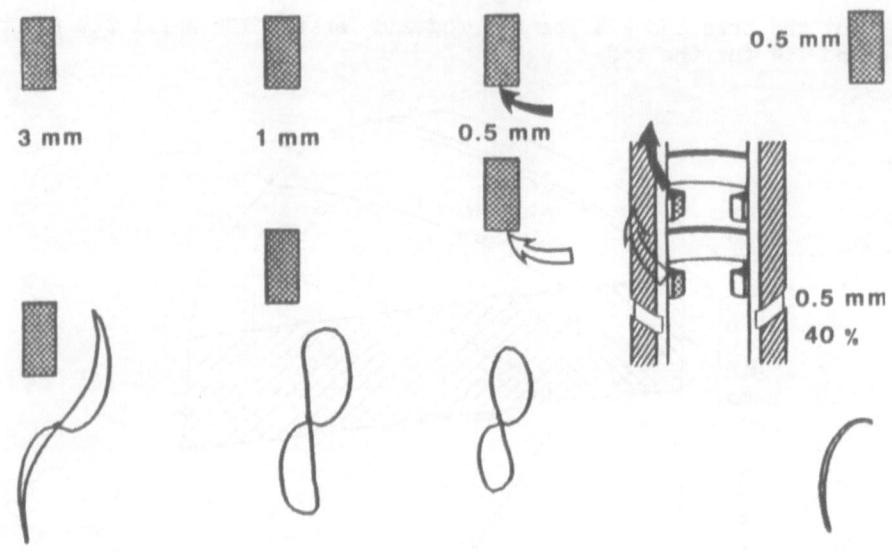

Fig. 4: Influence of the winding distance

The fact that the amplitude of a differential coil signal is not used for the determination of hte defect depth does not include that the amplitude is of no importance; the larger the amplitude and the more linear the relevant parts of the signal shape the more accurate the determination of the defect depth can be performed from the phase information.

One absolute and three differential coils excited with 400 kHz are moved along a circumferential slot on the outside. The four coils have the same widht of the windings but different distances between the two halfs when used as differential coil.

The signal locus curve at the lower right resulting from the absolute coil represents the maximal signal amplitude than can be achieved. In differential arrangement this amplitude (and its opposite) can be achieved if the distance between the two halfs is so large, that no interaction exists; then they behave as absolute coils. A distance of 3 mm - left signal - is an acceptable approach; but this signal does not have a well defined main phase direction which could be evaluated.

The more both parts of the coil approach, from 3 mm to 1 mm to 0.5 mm, the more the amplitudes of the loci curves reduce. The reason is that the second half of the coil already reacts upon the defect when the first half has not yet left its area. In terms of amplitude and linearity of the main signal direction the coil with 1 mm distance between its ports represents the optimum for the given inspection task.

3.3 REDUCTION OF THE AXIAL SENSITIVITY OF EDDY CURRENT COILS

The goal of these investigations was to reduce the axial sensitivity of eddy current coils and especially their dependency upon the frequency.

This task arises from the idea to reduce remaining signals when applying multifrequency-subtraction methods which originate from a different axial sensitivity of the applied frequencies. The tool used to predict the signals was again the 2-d Fe-code /2/. Representative for these investigations fig. 5 shows the 50 kHz signal loci curves for a copper-shielded (b) and an unshielded (a) absolute coil when it is moved along a circumferential slot, 0.5 mm wide, 60% wall reduction.

The left hand signals represent the real part, the right hand signals represent the imaginary part of the impedance variation. The effect of the shielding is considerably.

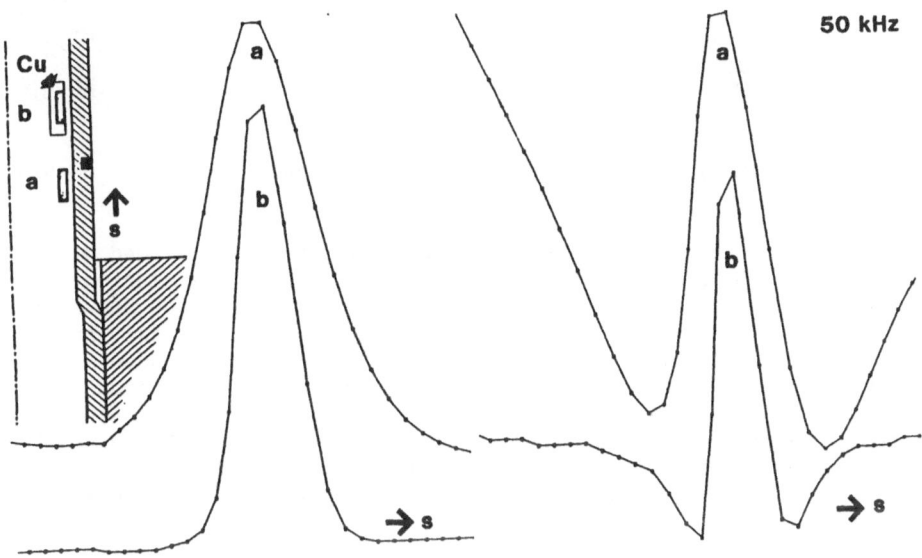

Fig. 5: Reduction of the interaction volume

3.4 SIGNAL INTERPRETATION IN STEAM GENERATOR TUBE INSPECTION

The following task is assumed: the signal locus curve on the right hand side of fig. 6 has been measured with an absolute bobbin coil at 25 kHz at the upper edge of the tube sheet. Can this signal originate from a deposit? If yes, what is the material and what are its dimensions?

254

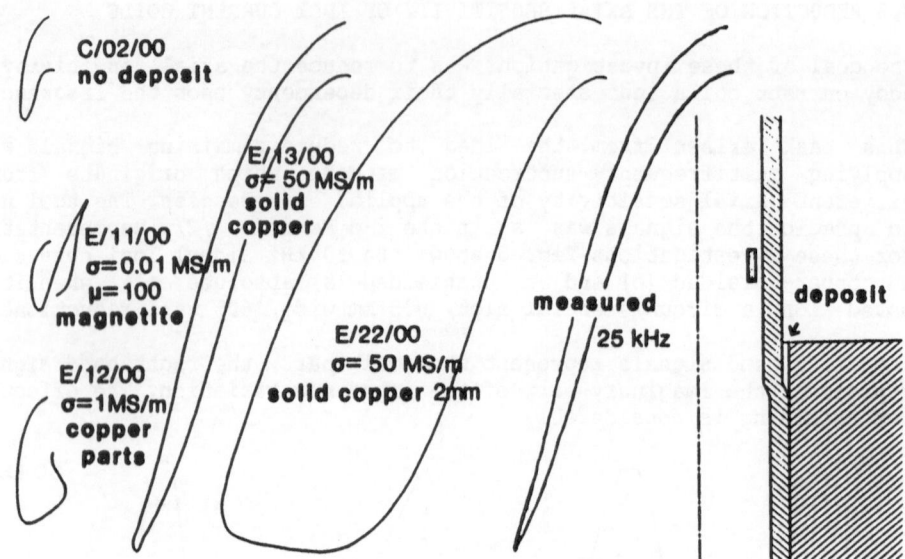

Fig. 6: Interpretation of eddy current signals in steam generator tubing

The task is solved as follows: the 2-d FE code is provided with the known input data of coil, tube, tube sheet and roll expansion. In the crevice gap between tube and tube sheet, three different deposits are simulated: solid copper, magnetic and nonmagnetic material having an electrical conductivity of 1 MS/m. The choice of this conductivity has two meanings: in its first meaning it represents a solid deposit with a conductivity of 1 MS/m. In its second meaning this conductivity represents a deposit having basically a much higher conductivity which occurs here as sludge or at least in small particles. Therefore the relevant factor that determines the conductivity is the resistance between its particles.

The 2-d code provides the signal loci curves in fig. 6 for the different deposits. The measured curve is simulated the best by E/13/00 which corresponds to solid copper just in the crevice gap. If the copper deposit would also cover the tube sheet the signals of E/22/00 would occur. Therefore the result is: the signal is produced by a - axisymmetric - deposit of solid copper, which does not exceed the upper edge of the tube sheet. This interpretation can be made redundant by evaluating several frequencies.

3.5 σ, μ-METER

Most of the eddy current techniques applied in steel-, machinery building- and automobile-industry for steel grading and microstructure characterization use encircling coils in a differential arrangement. These coils have an excellent resolution for measuring the permeability of ferromagnetic test pieces especially with low frequencies. - If a fill factor of 1 is assumed, the normalized inductive component indicates the figure of the permeability! -

Large extended components however have to be inspected with pick-up coils. The simple air-pick-up coil (only windings, no metallic or fer-rite components used) cannot provide this μ-resolution for ferromagne-tic materials. There is a μ-resolution in the range between 1 and about 10; but μ-values higher than 10 will converge in the normalized induc-tive component. The high frequency resolution (f ≥ 500 kHz) of these μ-values does not help, because these signals cannot be separated from those caused by σ-changes.

In order to overcome this problem a ferrite cup-core coil was used. Fig. 7 shows this coil in side view.

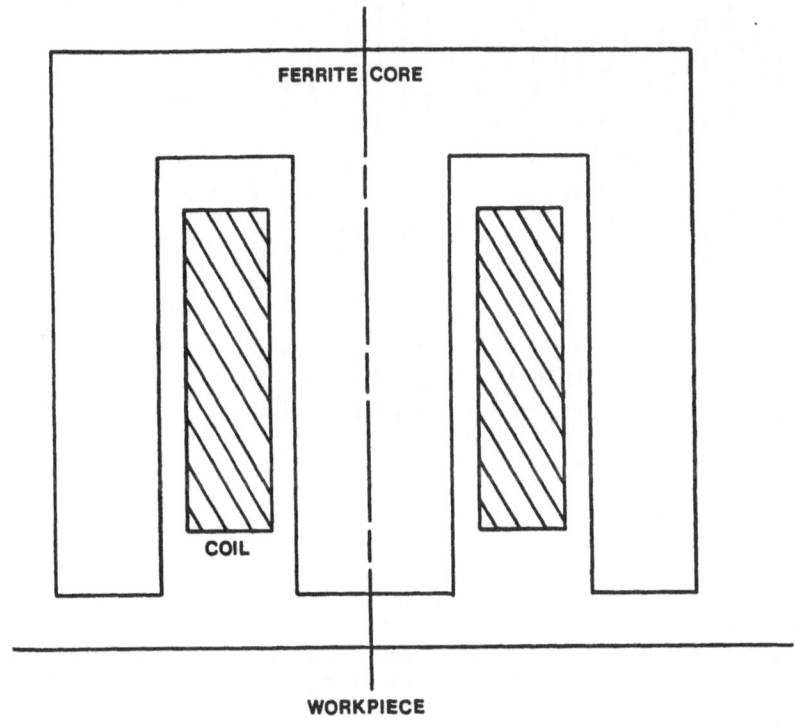

Fig. 7: Ferrite cup-core coil

Fig. 8 shows the corresponding 2-d-FE mesh generated to calculate the coil impedance and its change according to σ, μ-variations.

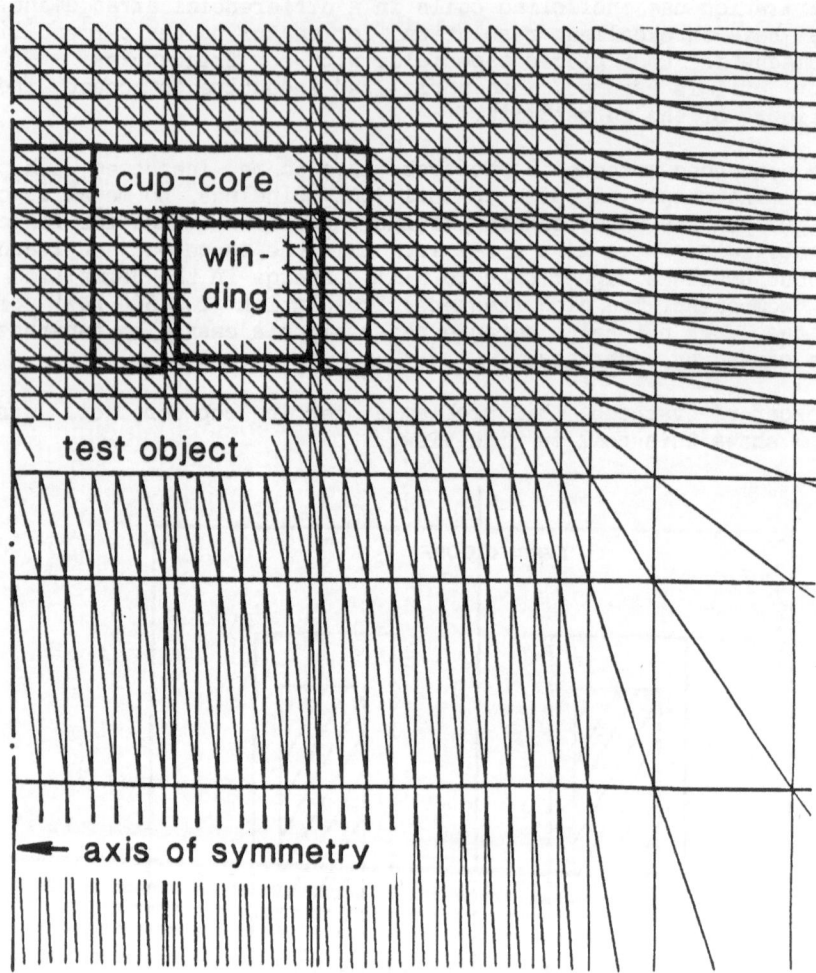

Fig. 8: FE-mesh of the ferrite cup-core coil

Fig. 9 demonstrates the advantage of the ferrite-cup-coil related to a
conventional air-pick-up-coil. In the upper part the normalized impedance for both types of coils is discussed for one steel specimen
($\sigma=3.7$ m/Ωmm^2, $\mu=86$) as function of the frequency. In the same frequency range the cup-coil has an impedance change which is almost five
times the change of the air-pick-coil. In the lower part of Fig. 9 both
coils are compared in order to estimate the sensitivity to separate
different steel samples. Even here the advantage of the cup-coil is observed. The reason for this fact is the good magnetic field conduction
of the cup-coil whereas a pick-up-air-coil produces large magnetic leakage fields in the air.

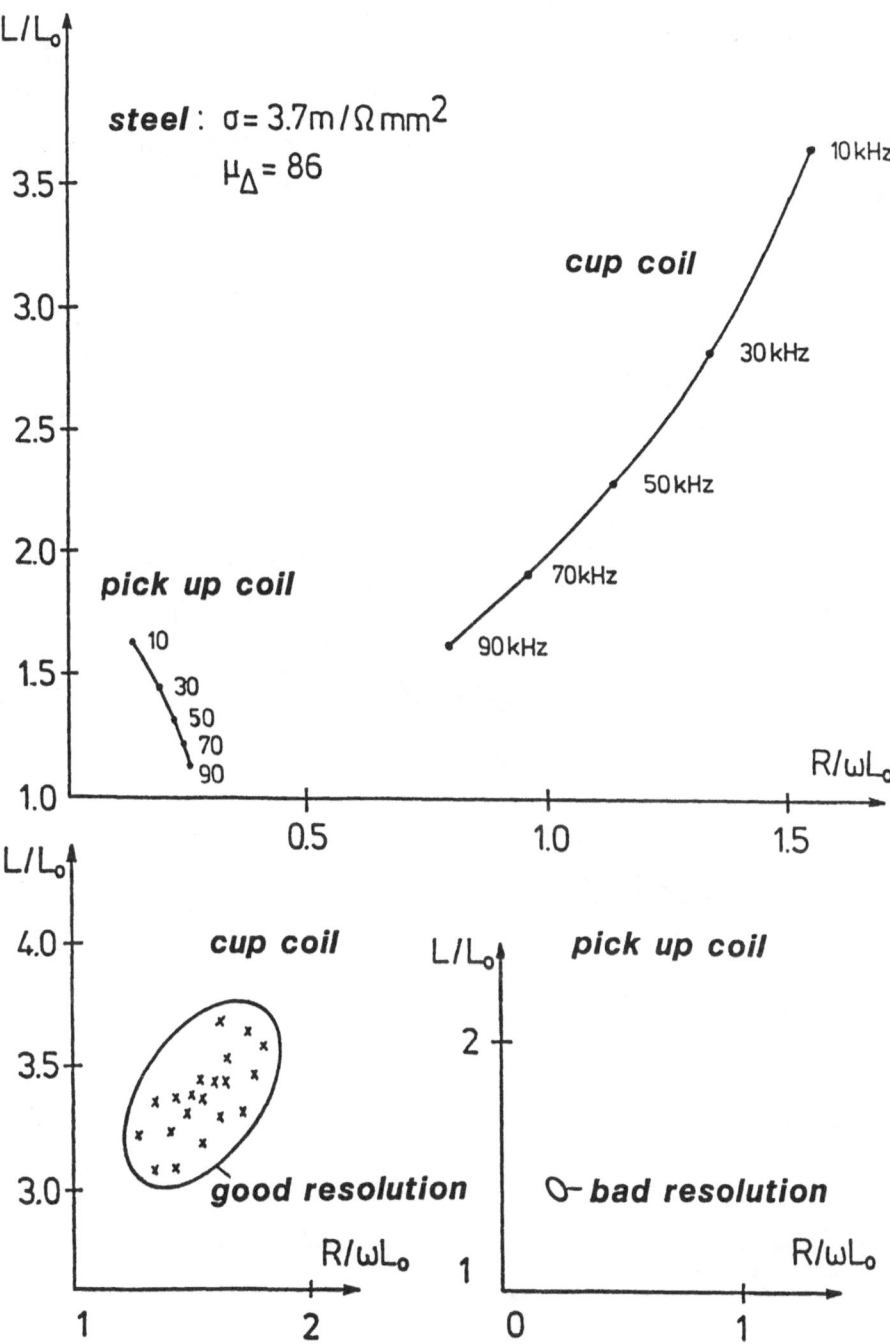

Fig. 9: Steel grading with different coil types

The finite element modeling has been resulted in a calibration of the cup coil impedance as function of (σ, μ)-changings for a frequency of 10 kHz. The transformation is a bi-unique-transformation. Fig. 10 shows the (σ, μ)-mesh in the impedance plane. The mesh is stored in a table in the storage of a microprocessor-controlled equipment which is called (σ, μ)-meter. The equipment measures the impedance of the cup-coil. By a table-look the (σ, μ)-values are estimated, at which the table values are interpolated.

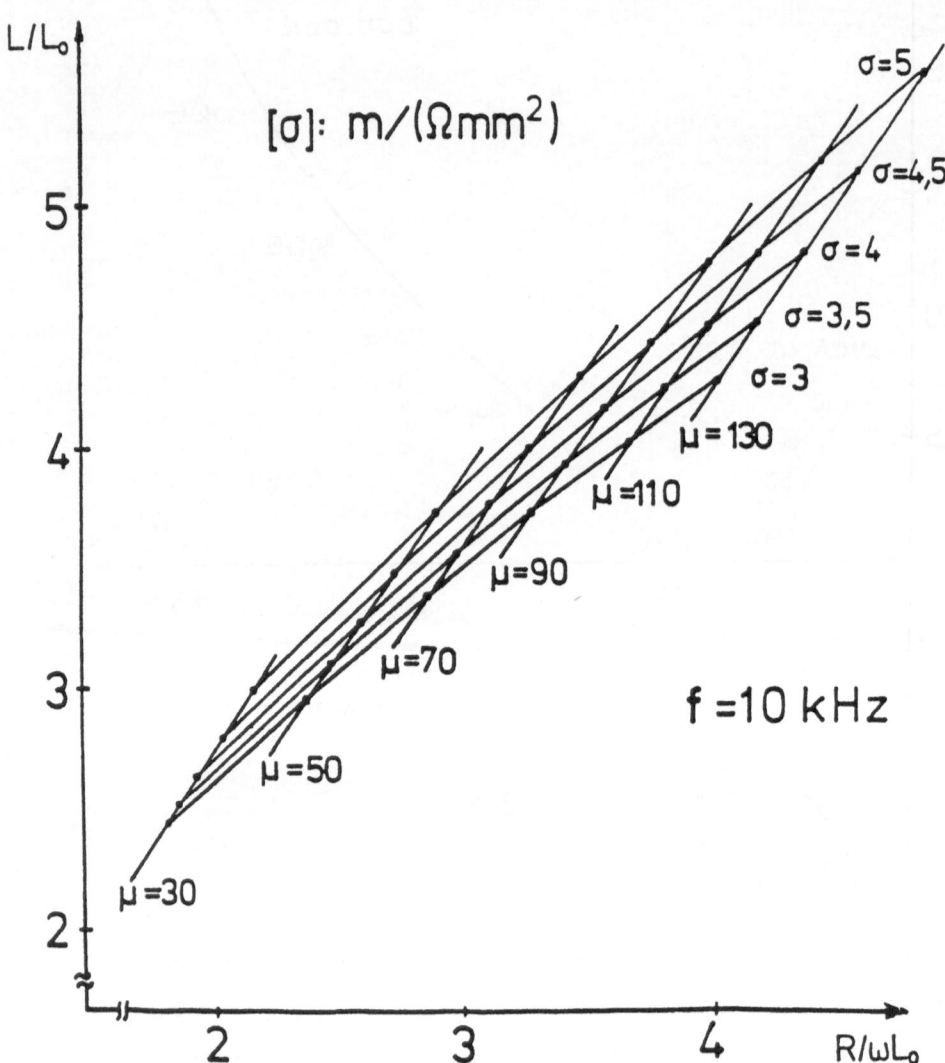

Fig. 10: Calculated σ, μ-array

Fig. 11 gives a result to a steel grading problem. Three different types of steel can be separated by the use of the (σ, μ)-meter in the (σ, μ)-plane. The scatter in the values represents the allowed scatter in the chemical composition.

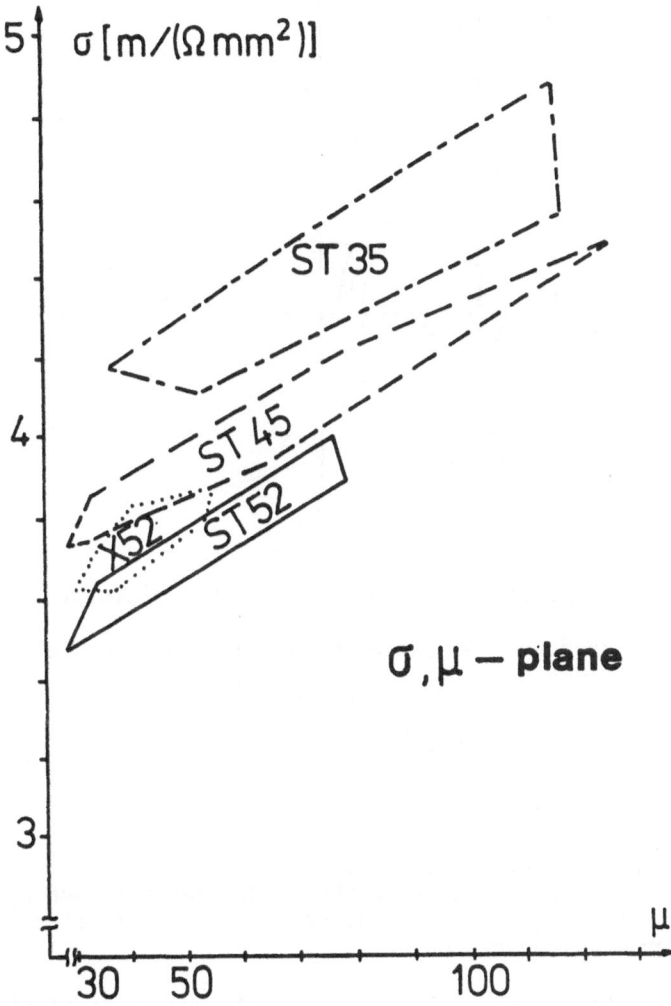

Fig. 11: Steel grading with a cup-core coil

3.6 SOLID BLOCK INSPECTION WITH PICK-UP COILS

Fig. 12 shows a real 3-dimensional problem for eddy current inspection.
It represents the top view on an austenitic block having a slot per-
pendicular to the surface. A pick-up coil is moved across the slot; in
Fig. 12 the coil is stopped in a symmetric position to the slot.

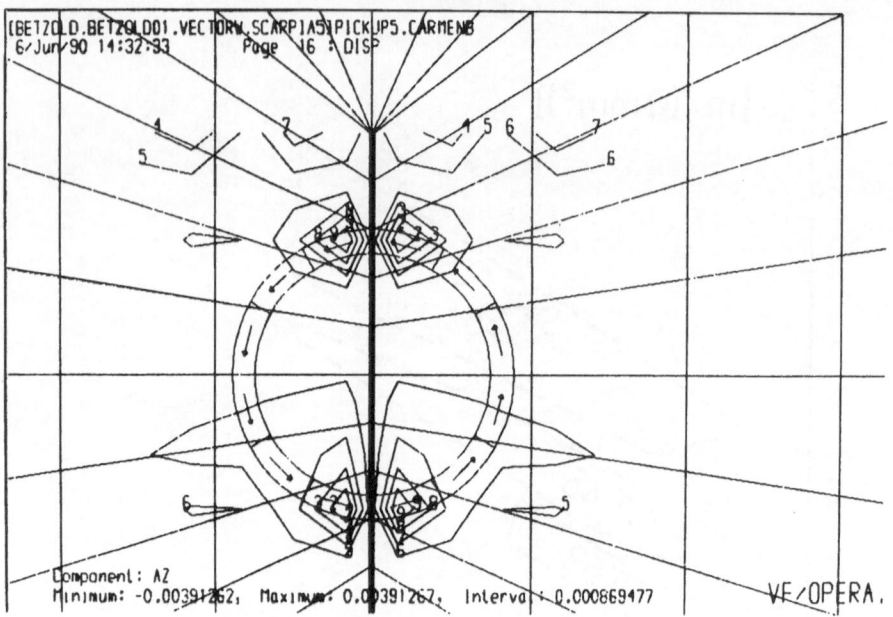

Fig. 12: Normal component of vector potential in the vicinity of a slot

The configuration has been modelled with the 3-d FE-code (ref. fig. 1
row 3) Carmen from Vector Fields /4/. The quadrilaterals in fig. 12
represent the element discretization of the inspection problem. As a
typical result, the normal component of the vector potential has been
displayed. In a defect-free problem the vector potential and the eddy
currents have only an azimuthal and no normal component.

If a defect is introduced, the eddy currents are deviated along the rim
of the defect, representing a new surface. Depending upon depth and
lateral extension of the defect, upon the electrical and magnetic properties of the material, upon the frequency and the coil size and arrangement additional components occur.

For the given inspection task of fig. 12 a normal component of the vector potential occurs on both sides of the slot entering the material on
one side and leaving it on the other side. This is demonstrated by the
lines of constant vector potential.

The change of the impedance due to the defect is computed from energy terms and available for signal loci curves. The potential for the validation of inspection procedures and for the interpretation of inspection signals is straight forward.

3.6 MAGNETIC PARTICLE INSPECTION TECHNIQUE

The magnetic particle inspection technique involves applying a magnetic field to the test piece either from magnets or by an electric current, with the latter being passed through the component under inspection. A suspension of fine ferromagnetic particles is then applied to the region of interest and the localised high magnetic fields which develop around a defect attract the mobile particles causing indications to appear. Magnetic particles require, however, a minimum magnetic field level and field gradient before they will respond to a defect. Since defects which produce leakage flux values below this threshold will not be detected by MPI, the reliability of the technique for the detection of small defects requires that precise information on leakage fields is available.

Computer modelling solves a twofold task:
First it is used to optimize the magnetization method for the selected component (single or multiple magnetization coil, single or multiple current contacts). For each configuration, the distribution of the magnetic field distribution in and around the component is computed.

The region under inspection has to be raised to magnetic fields of $\geq$ 2.4 kA/m ($\hat{=}$ 30 Oe) for weldable steels as specified in the "Handbook on the Magnetic examination of welds" from the International Institute of Welding (IIW) /6/.

The second task is to determine which defects (size, type, location, orientation) can be detected.

Fig. 13 shows the distribution of the modulus of the magnetic field distribution at the surface of a suspension part of a truck. The excitation is provided by a 50 Hz-magnetic field between the two ends of the hub (field direction indicated by the arrow). The scaling of the displayed values is in Oerstedt; the black missing area at the front of the hub represents field strengthes larger than 50 Oe.

Fig. 13: Magnetic fields at the surface of a suspension part

4. SUMMARY AND CONCLUSIONS

Three different categories of modelling tools for eddy current ndt have been presented - analytical, 2-d FE, 3-d FE - including characteristic applications for all of them. The results of the applications outline that modelling can contribute to at least 4 topics:
- optimized CAD of inspection problems
- validation of procedures
- interpretation of inspection signals
- education and training

The choice of the best tool depends upon the geometry of the test object, its material quality, the coil(s), the goal of the inspection (f.e. defects) and the disturbing influences. It is not at all true that 3-d modelling is the non-plus-ultra. The analysis of an inspection problem should start with checking the potential of the analytical model (exact, computing time in the range of seconds) and continue with 2-d models (approximation, computing time in the range of minutes). The appropriate modelling tools are available; the user has to apply them in the best efficient and economic way.

5. REFERENCES

/1/ C.V. Dodd
Integral solutions to some eddy current problems. International
Journal of Nondestructive Testing, $\underline{1}$ (1969) pp. 29-90

/2/ R. Palanisamy, W. Lord
Finite Element Analysis of Eddy Current Phenomena, Materials Eva-
luation,
October 1980

/3/ N. Ida, K. Betzold, W. Lord
Finite element modelling of absolute eddy current probe signals.
Journal of Nondestructive Evaluation 3, 1982

/4/ C. Emson, W. Trowbridge
Transient 3-d eddy currents using modified magnetic vector potenti-
als and magnetic scalar potentials
IEEE Transactions on magnetics Vol 24 Nr.1 Jan 1988

/5/ J.M. Coffey
Mathematical modelling in ndt - what it is and what it does
in M. Blakemore, G. Georgiou
Mathematical modelling in nondestructive testing,
Clarendon Press, Oxford 1988

/6/ Handbook on the magnetic examination of welds
The International Institute of welding Document IIS/IIW -849-87